VERTICAL LIFT

A Heli-Pilot's Mission for Freedom, Family, and Success

George Kelham

As told to K.H. Bickell

George Kelham
slheli.com

Written (as told to) and co-authored by K.H. Bickell
Editing and Interior Design by K.H. Bickell
Cover Design by Lucy Fox

ISBN: 978-1-7780650-0-2 (softcover)
ISBN: 978-1-7780650-1-9 (electronic)

DEDICATION

For my sons, Boyd and Taylor,
And for my grandson, Tucker,

So that they might know stories left untold.

CONTENTS

PROLOGUE

SOUTHAMPTON

July 19th, 1947

We do not know the way this story starts. Not for sure. Many of the people who were there are long lost to history, the facts of their once-vibrant lives changed from clear-scope technicolor to an image as hazy as an archived photograph, details veiled in sepia, black, and white. The scenes within this book have been cobbled together with tempered artistic license in order to honor both the craft of storytelling, and the real-life experiences the stories are based on. The opinions presented, and perspectives in which the stories are told, are based upon the experiences of the man who lived them and are meant to reflect his beliefs and attitudes at each varying stage of maturity. In some cases, names of people and businesses have been changed to protect the privacy of those involved or shield the identities of guilty parties, and greater creative liberty has been wielded within the first chapters of this book, as time tends to diminish our earliest memories first.

In all instances, though, this book is based closely on historical, lived events, utilizing research pulled from first person accounts, journalistic interviews with peoples flung worldwide, books and other sources dedicated to relevant matters, and records compiled by various governments and churches. These items have been mindfully collected and presented in hopes to both bring the past into greater focus and to entertain the reader of this text. And while we do not know all the details of every moment in time, what we do know are our own personal legends, and a family mythos of how it all started with spontaneous choice: fate, in disguise. From that moment onward, we can use our imagination.

We can conjure in our minds the sights and sounds of the famed port of Southampton; its cranes, dock sheds, freight boats, and tugs that crowded the seascape, seemingly as populous as the gulls that perhaps a toddler laughed at in delight, his own small arms waving in mimic of their wings. Ocean liners likely casted shadows on the harbor's boardwalks, temporarily blotting the summer sun and causing a frenzy of diminutive tugboats, each racing to secure their behemoth and bring it to berth. We can feel within our own sinew the same tired muscles; arms that had pushed the entirety of a young couple's belongings on a small trolley through a pulsing crowd of other would-be immigrants, fellow Brits in mass diaspora following the country's bleak war years. We can feel the relief of setting oneself down upon an empty wooden bench, and we can imagine the forthcoming weariness of the nearly month-long journey on which this family, already exhausted by the bustle of the port, was about to embark.

"You know," we might imagine the patriarch mused, "they say regular folks will travel by air within our lifetime. Think of that. A person could cross the Atlantic in a matter of hours. Or hop on a rotorcraft to lunch in another county." We imagine he whistled then, a sound that started high and ended low. Perhaps his wife steeled herself against a shiver brought on by the thought. But for the V-1 bombers that shook the underground tunnels she'd sheltered in, the young woman hadn't much experience with aircraft.

"I don't know about that." Perhaps she shook her head or beamed the beatific smile she was known for. "But I think I'd rather stay out of the sky."

It was July 19th, 1947. While the weather was likely fair, the wait was very long. John Leonard and Louise Susannah probably split their attention between wee Francis John and the giant ships that came into slow but steady focus as they neared. The couple watched that Francis stayed close, kept out of the way of other travelers, and remained a good distance from the barrier, lest he slip through and fall into the water below. The boy was only two years old, an age in which children operate with an energy their parents struggle to remember ever having had themselves. Occasionally his parents' eyes might have darted from the little boy to the sides of the arriving liners, squinting for the name recognizable as the one they were meant to board, their hearts thrilling at first sight of each new steamship and sinking with disappointment as the letters branded upon them came into readable distance and proved not theirs. They had waited for hours.

The family was on their way to America. The first stop would be New York, with the ultimate goal of starting a life in Alaska - two very different places, but both entirely foreign and understood to be bitterly cold. John Leonard, as reward for serving the United States Navy as a commissioned officer, seconded from the Royal British Navy, had been promised hassle-free immigration to America and an easy route to citizenship once there.

It was unusual for a British Officer to serve a foreign power, but the Second World War had turned usual on its head in every way. Britain, the major force leading chief Allied powers, had the support of Canada, Australia, South Africa, and many more nations, but the Nazi-led Axis union had proven unrelenting. Ship after ship of Allied forces were sunk and the cause had become desperate. Despite multiple campaigns delivered by much of Europe and the world, America had withheld concrete support until after Japan's attack on Pearl Harbor in December 1941. Finally rallying their troops to bring end to the chaos that had, until that point, taken place entirely off their soil, the United States government discovered that while their country had produced an abundance of ships, equipment, weapons, and vehicles to support war efforts, they sorely lacked expert seamen. The proposed agreement was thus: should officers of the Royal British Navy volunteer to man American ships - falling under American command - they would be offered full entry into, and right to take permanent residence within, the United States of America.

John Leonard was one of many sub-lieutenants who had been granted transfer from the Royal British Navy in order to assist America's aligning campaign. A serious and reserved man, his professionalism, work ethic, and dutiful service had earned both a swift rise through rank and invitation to serve as an officer for an additional year after fire officially ceased in 1945. He'd spent 1946 patrolling in Japan while his resourceful Louise Susannah operated a small bed and breakfast, all the while tending to the couple's young son. Together, they had saved just enough money to make the dream of immigration a plausible reality.

Not that anything was certain. While there might have been a few more pennies in the purse than usual thanks to the couple's hard work, John knew how quickly a growing family could make use of them, even despite Louise's frugal nature. But if the last decade had taught John anything, it was that a man had to both follow his instincts and seize opportunity. After all, immigration was not the only right extended to John by the United States Navy. He had also been given the privilege of temporarily opting out of duty as he saw fit during war years. This was a right John had flexed only once, and the reason why is lost to time. Perhaps he awoke with a bad stomach or was required to sign legal paperwork that could only be fulfilled within a certain timeframe. Or maybe Louise had asked him to stay home, just one more day, just that once. Alternatively, he could have had a gut feeling. Whatever the reason, John had only once declined to board a ship to which he was assigned, choosing, instead, to board a later vessel. Within days of its departure, the original liner had been sunk by a German battleship. There had been no survivors.

Taking advantage of the opportunity extended to him – even just the once – had spared John's life. Perhaps he imagined that taking advantage of America's other offer might just change it.

For her part, Louise Susannah was a woman who valued fresh starts. Born Veronica Claire Floate, she'd changed her name to Louise Susannah Floate upon converting to Catholicism before marrying John Leonard, who was a practicing Catholic. The tradition of changing one's name upon Catholic conversion was antiquated even then, but Louise Susannah was not the kind of person to do anything by half measures. Obviously, she saw the potential that immigration held for the family. England was a place where one's station in life was very much set, and though John had enjoyed success as a naval officer, post-war his professional opportunities were limited to that of a humble fitter-and-turner. The United States were said to be a place where one determined their own station, regardless of working-class birth. Still, we can wonder if perhaps there was something about the severity of independent America that troubled Louise. America wasn't only foreign to the county of her birth and every ancestor in her family line, but it wasn't even Commonwealth. And, for as forward-thinking as Louise was, we also know how deeply she valued tradition, origin, and one's roots. After all, while changing her very identity was a mindful symbol of spiritual transformation, the names she'd chosen to adopt were that of her own mother and grandmother's; homages to lives once lived.

But likely, there was more to Louise's decision to leave. Likely, a sort of grief played its role.

While John hailed from Guilford, Louise was a born-and-raised true Cockney; her whole life spent within hearing distance of the Bow Bells. Their famous chimes carried through the district on a southwestern wind that persistently blew for more than three thirds of the year, and, though their ringing must have become nearly unnoticeable by their familiarity, the bells had sounded Louise's life's song. Their melodious clanging was a point of comfort. They were a touchstone, a guarantee that in a rapidly changing world some things still stayed the same, some things could still be counted upon. Their sound had come to epitomize security, safety, consistency. Home.

But the bells had silenced in 1940 as part of Britain's anti-invasion preparations, and Louise had not heard them ring since. Only months later, just before she and John wed in 1941, German bombs had reigned down upon the city, demolishing the Bow Bells completely. That night, Louise Susannah had forced her breath to remain steady as she sat still and silent, taking precarious shelter underground, crammed shoulder to shoulder with frightened neighbors. Perhaps she closed her eyes to ignore the way the lamp light flickered in response to the destruction above, or to avoid witnessing the terror on the faces of children she'd watched play on the street outside her front steps. We don't know if it was that same night, or another, but along with Cockney's symbol of home, Louise and John's actual house had been reduced to rubble. And while they'd later made happy memories in a new place on Carleton Avenue, where Louise let out an extra room while John was on his last tour of service, those remembrances didn't include the sounds of the Bow Bells and so, perhaps, felt nothing like home. Perhaps London, with its missing buildings and people and bells, already felt like a new country to Louise. Perhaps the sight of the grey, rubble-reduced city, though in rebuild, had continued to feel as though Louise had stepped blind and blinking from the dark underground into a strange and ruined place.

"Here we are," we imagine John saying, breaking the lull of his wife's daydream to draw her attention to an arriving ship.

"Let's hope this is us."

Perhaps Louise turned her head from where Francis played on the boardwalk to the sparkling, blue water as a mighty ocean liner sailing into view.

"The Indrapoera," the couple might have heard a port worker say, behind them. Perhaps they felt a flash of disappointment that the ship wasn't theirs, but it couldn't have lasted long. The vessel was too beautiful, too mighty, too impressive: it's elegant black funnel giving way to a dove-grey and white hull with a vertical bow, and open, airy docks, above which the Dutch flag waved, as if in celebratory greeting, simply a different combination of America's similar red, white, and blue.

Had John and Louise overheard port workers conversing, perhaps they would have learned that this charming ship, with her exotic, evocative name and mysterious, oriental overtones, was known as *one of the graceful daughters of those Ruys gentlemen*; part of an impressive fleet known for its luxurious offerings, owned by the Rotterdam Lloyd company. The Indrapoera, like John, had served in the war and had, coincidently, also been spared due to a happening of similar chance. On April 27, 1941, the Indrapoera was meant to accompany a sister ship, the Slamat, who set sail with six hundred troops aboard out of Nauplion. The Indrapoera, however, had suffered engine problems at short notice and was forced to stay in safe harbor. Shortly thereafter departure, the Slamat was bombed mercilessly by a German fleet. Of a total of a thousand men, only fifty survived, the remainder finding watery graves in the Aegean Sea. Had the Indrapoera not suffered her engine problems, she might well have met the same fate: such, John knew too well, were the fortunes of war.

"Excuse me," Louise might have politely asked the portmen conversing behind her. "Where does the Indrapoera sail to?"

"Cape Town," a dock worker might have replied. "If you want on, Missus, go on then. All Brits are welcome to set up house in South Africa. Government relocation allowances provided and all."

Perhaps the tired couple looked at one another, then. The virtues of Imperial Africa were commonly exalted in British news and literature, lyrically romanticized by the likes of Rudyard Kipling, Ernest Hemingway, and Karen Blixen to be a never-ending paradise of abundance and adventure. The people of the English-settled colonies of South Africa and Rhodesia, too, had been celebrated for the part they had played in Allied efforts. Self-governing Southern Rhodesia, despite its small population and lack of diplomatic power, had contributed more soldiers per capita than any other nation in the world, and the man force, made up of all the races that resided within the colony, had proven themselves as courageous, humble, and dutiful warriors.

Perhaps Louise thought of the address young Princess Elizabeth had given on her twenty-first birthday, only three months prior.

As I speak to you today from Cape Town, I am six thousand miles from the country where I was born. But I am certainly not six thousand miles from home. Everywhere I have travelled in these lovely lands of South Africa and Rhodesia, my parents, my sister, and I have been taken to the heart of their people and made to feel that we are just as much at home here as if we had lived among them all our lives.

It wasn't England, but it was Empire. Commonwealth. A fresh shoot nurtured by established roots. A new identity, but a family name.

"They say it's awfully cold in America," John Leonard might have said, raising a brow.

Perhaps Louise's beautiful smile beamed again. "Sun would be nice."

Maybe then they looked into each other's eyes, searching one another for courage, for daring, for game. *Warmth,* they may have collectively thought. *Freedom. Adventure. Opportunity.*

If not bells, then birdsong.

"Let's go," Louise Susannah might have said.

And then, we can imagine the weight of a small child hefted to one's hip. Of the rickety trolley, again, pushed into motion. Of the feel of thick paper tickets between one's thumb and index finger. Of Louise's pen scratching the passenger list: name, occupation, age.

KELHAM, John Leonard. Mechanic. 32
KELHAM, Louise Susannah. Housewife. 32
KELHAM, Francis John. 2

Country of Intent: South Africa.

PART ONE

UMTALI

May 29th, 1966

Eleven-year-old George Kelham knew he'd overslept the minute he opened his eyes. The whitewashed walls of the big square bedroom his father had built onto the back of the family's small, boxy house were lit too brightly by the sun for it to be the usual time he woke up on a Sunday morning. George felt a fleeting rush of worry that his parents would be annoyed he was still in bed so late, but quickly realized that they must have let him sleep in on purpose. After all, it was his mother who got everyone up in time for a healthy breakfast before church. Mom was a health fanatic, and always made sure the family started their day off right with freshly picked fruits and vegetables, most often accompanied by porridge, toast, and poached eggs. Mom had strong, white teeth which she credited to good nutrition, reminding her children often that most people who'd grown up in England had very poor dental health because they ate mostly starchy, sugary, dense foods that had had all the goodness either boiled or pickled out of them. It was hard to find good, fresh food in England, but in Africa, it was easy.

George turned his head to see the bed of his eldest brother, Francis, and tried to make sense as to why the home wasn't its usual hive of activity. By this time, George should have been able to hear all members of the family readying themselves for mass, where George, an altar boy, was counted on to perform such important duties as ringing bells, carrying incense, and lighting candles. George found the job pretty boring, but the priest was nice enough – unlike the heavy-handed Brothers that taught at George's school – and at least it was something to do other than sit still in a pew. But Francis's bed was empty, its sheets and blanket still tucked with the tight precision his mandatory military conscription had required years earlier. Francis had attended a friend's bachelor party the evening before and must have stayed out all night.

Dad won't like that, George thought, even though the idea of Dad being angry over Francis' stopping out was a bit rich. Dad went out for beers with friends every week and often came home smashed, waking up the whole house by picking a fight with Mom in the small hours of the morning. Besides, at twenty years old, Francis was technically a grown-up, even if he did still live at home. He had a good job at the post office but was saving for a trip to Europe. Once he left, George would have the big bedroom all to himself.

George had been waiting for his very own bedroom all his life. The youngest of four children, he'd always shared the back room with his brothers. Originally, the home on Number Fourteen, Rudland Avenue, had featured only two bedrooms, but Dad had built the addition after Marion, the only girl in the bunch, was born. Then Mom and Dad shared the master, Marion got the smaller second bedroom, and the boys shared the newer, larger space. Seventeen-year-old Robert, second born after Francis, had recently moved out of the family home after having secured a job with the Ministry of Water. Now Robert lived in a workers' camp about three hours away. That left only Francis to go before George's dream came true.

But while the idea of having his own bedroom made George very happy, it also made him sad to think of Francis moving so far away. England was exactly twelve thousand six hundred and forty-one kilometers from Rhodesia, and that was very far – almost twenty hours by plane. Francis was born in England but had told George he didn't remember it there. He and Mom and Dad came to Africa on a ship when they emigrated. First, they'd arrived in South Africa, but were made uneasy by the popularity of the authorial National Party and its increasingly rigid apartheid laws. Instead of staying there, they'd bought a vehicle and driven to the small city of Umtali, Rhodesia, where the family set roots. That journey had taken a *lot* longer than twenty hours.

George's thoughts were interrupted by a sound from the kitchen. He lay still, quietening his breath, trying to hear his parents' conversation. His mother and father were speaking too low and quiet for George to make out proper words, but their voices sounded strange, like someone had a sore throat. When George heard his mother sniffle, he realized she must have woken up with a head cold. Perhaps that was why the family had taken the morning off from weekly mass. George would never wish for his mother to be sick, but he smiled as he imagined the afternoon that suddenly stretched wide open before him like the long, green yard outside his bedroom window, his mind delighting in thoughts of how he might spend an extra unplanned day.

It was of young George's opinion that most of the people in his family simply didn't know how to enjoy themselves. When he was younger, he spent most of his time playing with Marion, who was closest to him in age and used to be good for a laugh, but now his sister was thirteen and only interested in girly things which George, emphatically, was not. Mom was always busy tending to the house or garden or mending clothes, and while she was loving and affectionate and usually in good humor, she didn't know how to play, either. Robert – a boy, and only six years older than George – might have been more fun, but he had always been serious and studious and just like Dad with his nose forever stuck in a book. And Dad… well Dad was disappointed in George most of the time, a fact he made no attempt to hide. George knew Dad wished that he were more like Robert but, try as he might, George just couldn't bring himself to care about homework or grades or really anything about school at all.

What George cared about was adventure. Adventure, and action, and speed. George didn't want to sit around reading or thinking about nonsense he'd never use or watching the day pass by outside his window. He wanted to go somewhere, and he wanted to go there fast. He wanted to see new things and feel the way his stomach leaped and made laughter holler from his mouth like when he and Jimmy Hynes raced homemade go-carts down the hilly street in front of their homes.

Jimmy was George's very best friend and lived only a short walk away. During summer break, George would wake early and walk to Jimmy's immediately after eating breakfast. He'd sit in the Hynes' family's back garden and wait for his friend to wake up, and then they'd spend the whole day together: hunting snakes in the jungle or climbing down through manholes where they could navigate through the whole city underground, sometimes popping up through the sewer system to find themselves inside an empty factory, which they'd then explore to see what fun they could find within. Sometimes the boys would spend their evenings at the cinema watching Clint Eastwood in spaghetti westerns, or they'd go "roof rattling," crouching in the trees by neighboring homes and tossing pebbles onto tin roofs, muffling giggles with their hands as the rocks noisily slid down the metal thatching, fleeing when annoyed homeowners stomped out in search of the mischievous culprits. Guy Fawkes Day was also a banner occasion. All year long, Jimmy and George would save up pocket money and search for used Coke or Fanta bottles to exchange at the gas station for the thruppence. At the end of October, they'd go to the Indian market, which imported a dazzling display of explosives, and the boys would spend every pound they had on fireworks, rockets, and blue-fused flash bombs, which they'd cram into every crevice of the huge bonfire they'd built and would add to for days. Then, on November fifth, Jimmy would light a match and blow the whole thing to smithereens.

But even simply hanging around the boisterous, welcoming Hynes' family home could be fun. Their house was a stark contrast to George's, which always seemed a little too cold in feeling, a little too reserved, and little too quiet to be happy. At Jimmy's house, modest as it was, there was always music and laughter and a smile or wink from Jimmy's parents, who enjoyed chatting with their children and their children's friends. Mr. Hynes was a train conductor and, like George's father, had served in World War Two. He had originally hailed from Manchester but had been stationed in North Africa during his air force service. When the war had ended and he'd returned to England, Mr. Hynes had noticed a help wanted ad that invited British men to apply for jobs building the Rhodesian railroad. The Hynes family had emigrated to Africa the same year as the Kelhams.

The Hynes children, Angela, Jimmy, and Brian, were famous for their adopted menagerie of local wildlife and their generous parents had come to tolerate all sorts of small, found creatures. There were pet hedgehogs, rats, and snakes to play with. Even buckets of small fish were pulled from the river and left in the corners of common areas. George and Jimmy, the latter enamored by the endlessly glamorous lifestyle of young adults and the former enamored by Jimmy's teenage sister, whom George would often gift small presents to in an effort to impress, made an almost daily mission of spying on Angela and her friends, which sometimes included Francis, as George's elder brother and Angela were occasionally a bit of an item. Seeing his elder brother in the company of Angela Hynes – in the Hynes' wonderful home, no less – made George's chest swell with pride.

In fact, it seemed like George and Francis were the only members of the Kelham family who knew how to live at all, which was why Francis was George's favorite, and was why George would miss Francis so much when he was gone. Once Francis was in England, there'd be no more spontaneous visits over the border to Villa Delamonica where Francis would sometimes take George for ice cream or for a coke on his days off. George was always pleased as punch to go wherever his funny, friendly big brother offered to take him. He loved being by Francis's side, watching him charm the prettiest girls, tell jokes with his friends, or seeing the adults fawn when he returned from a day trip with demijohns of wine to share, alcohol being expensive in Southern Rhodesia for how heavily it was taxed. Adventuring with Francis gave George a glimpse of what life might be like once he was old enough to live it for himself, and without that glimpse, a good life could be hard to imagine.

It seemed there were so many things a boy had to do before he was allowed to make his own choices. So much of George's time was already accounted for, with little thought of how he might want to spend it, or what he might want to learn. There were boarding schools far away from friends and family, grueling sports made to exhaust him, duties assigned to fill every spare hour, and church responsibilities piled on even higher, just in case he found a minute to think for himself. And then there were more rules – not always laws, like there were in South Africa – but unwritten social policies that kept everyone in their place. In Rhodesia, for example, a black person could technically go to the same movie theatre as a white person, but they wouldn't because it just wasn't done, and probably because they couldn't afford to even if it were done, and if they could afford to, they might be made to feel uncomfortable. To a lesser but similar extent, white people were kept divided by class. Richer children were discouraged from speaking to George, whose family was lowly working class, and although Mom and Dad paid a dear sum for

Marion to go to a fancy school where she had to wear white gloves, the higher status girls there would never dream of visiting the Kelham home, nor would their fathers ever share even a casual conversation with Dad. And even ignoring the unwritten laws of society and the regulations that came with school, once a young man had the opportunity to chase his own dreams, the government pulled him once more into a life of rigidly scheduled time and unquestionable rules: every white adult male in Rhodesia was required to complete at least nine months of military service before they were finally free.

The thought of it all was overwhelming and blocked any idea of the future, looming in the distance like some insurmountable wall George would have to scale, some terrible rite of passage that could deny George manhood, adulthood, any sort of agency if failed. George hadn't barely begun any of that slog and still had so many years to go, but at least he could see that Francis had gone through it all and had come out on the other side fit, smart, and able to make big decisions. Because of this, George could see that someday life would be his to lead. There would be adventures and travel, risks to take, money to make, respect to be won. When George finally became an adult, he'd be just like Francis. He wouldn't waste his evenings like Dad, sitting around letting drink dissolve his days into dull meanness. George would be brave, and in control, and always up to go on a trip or see a film or go out and listen to music with friends. There would be no more teachers looking down their noses at him, that was for sure. No more of Dad's smarting smacks to the back of the head. No more classmates thinking they were better just because their father's bank account balance was greater than his, and no more anybody barking orders. Francis was George's proof: he was so close to freedom and, someday, George would be, too.

But to be fair, Francis was also the reason that the Kelham children had to go to boarding school in the first place. As a young teen, Francis and a few local boys had broken into a record shop and were found by the police chucking Elvis Presley records off the building's roof. Mom and Dad were mortified. They went to the church for advice, and Bishop Donal Lamont suggested that Francis be sent away to study in South Africa, where he could receive a proper education with boys of better social standing than the petty criminals he'd been found in company of. Wanting the best for all their children, Mom and Dad decided that meant

Robert, Marion, and eventually George, would have to go to private school, too.

The year prior, in fact, George had attended the same South African boarding school with Robert. They had travelled there together by train, a trip that had taken two days. But just before his graduation, Robert had sneaked out of his dormitory and been caught smoking downtown with friends. This had resulted in Robert's expulsion, and because George was too young to travel so far by himself, it meant that both boys were brought home together. After their return to Umtali, sixteen-year-old Robert found decent-paying work and George was enrolled at Carmel College, the local boys' Catholic school. Mom and Dad had decided however, that George would be old enough to travel by himself once he turned twelve, and so it was planned that George would soon be sent away again, this time to a French-Canadian school in Que Que. George hated the thought of it. He wanted to stay at Carmel College with Jimmy.

Not that Carmel was perfect. Ran by the church, corporal punishment was swift and sometimes brutal. Jimmy was especially talented at attracting the negative attention of the crueler Brothers who thought nothing of battering children with their fists, or even kicking the wind out of their students as they collapsed to the floor. George mostly avoided such extreme punishment but had been caned a couple of weeks after school picture day, when the headmaster had opened mail from the photographer to find images that displayed George as the only child to make a silly face among rows of better-behaved, smiling children. But not all the teachers at Carmel were violent, and the philosophy of the school itself was staggeringly democratic in contrast to that of the Rhodesian government. At Carmel College, there was no segregation on admission, and its new halls – still under construction the year George and Jimmy began attending – welcomed students of all ethnicities: Black, White, Indian, Asian. To George, the best thing about Carmel College was that Jimmy went there, and the second-best thing was that George could sleep in his own bed every night as the school was within walking distance.

Built right on the edge of town, farms that neighbored the school were sometimes attacked by baboons, who would descend like locusts, consuming every living plant, fruit, and vegetable, ripping apart livestock and pets, and, sometimes, even unlucky humans. Perhaps because of this, Jimmy's mother usually drove the boys to school, but sometimes George chose to walk, and sometimes he woke up late, and then he had no choice in the matter at all. But while it was a long walk to school, it was *merely* a walk, and George didn't at all mind if his legs got tired. He was tough and liked being outdoors by himself, with no one telling him what to do. Sometimes, if he became hungry while walking, George might pinch an avocado or an orange or a banana from a tree along the route. Then he'd feel almost rugged, like he was really taking care of himself, master of the world. He imagined that was what his life would always feel like once he grew up.

Stomach rumbling in want of breakfast, George scrambled from his bed, rubbing his arms to warm them against the cool air that gave him goosebumps once out from beneath his warm covers. It didn't get very cold in Rhodesia, not the way Mom and Dad told him it did in England, where they might wake up to see that frost had turned the grey streets white, or that snow had fallen from the sky, or find that ice had started to form over water. But May was the very start of Rhodesia's rainy season and the day could start off as cool as six degrees Celsius, despite afternoon temperatures often rising to a comfortable twenty degrees, or even a blistering plus thirty. Quickly, George changed from his pajamas into shorts and an undershirt, slipping his feet into socks and shoes before pulling a secondhand sweater from his closet to wear until the day warmed up.

Before leaving his bedroom, George looked out the window, smiling in the certainty of fun that seemed almost promised by the clear blue sky. The world was alive and colorful and ready for him to enjoy. The little Kelham house sat on a half-acre of land, as did each of the neighborhood's almost identical little homes, rapidly developed to house the influx of British immigrants that arrived en masse postwar. Outside George's window, he could see his mother's vegetable garden and a large mulberry tree; its generous limbs stretched out wide like a thick green cloud enveloping its small but sturdy tree trunk. The Kelham children would help themselves to fistfuls of the tree's sweet fruit through most of the year, its dark berries staining the bottom of their shoes as they snacked. At the very back of the property, there was a smaller building where the houseboy, Phineas, lived. Phineas was an African boy about Francis' age and was hired to help Mom around the house and in the kitchen. He chopped vegetables and mopped floors and was sometimes even asked to keep an eye on George. Mom called the little house Phineas lived in "the servant's quarters," but that made the Kelhams sound much richer than they actually were.

Really, the Kelham family wasn't well-off at all, and George blamed school. Before Francis and Robert started private school, there had been enough money for Mom to take the older boys on a trip back to England. But then Marion was born, and then George, and then the Bishop suggested the children start going to expensive schools, and then there wasn't money for anything but tuition. The family lived without a television or a phone. George's father had driven the same modest vehicle since the family had disembarked from Cape Town, and George believed he'd probably drive it until the wheels fell off. George's clothes weren't only secondhand, but thirdhand as they first went through Francis *and* Robert, and Mom allotted the family only three days a week to bathe, in order to not stress the hot water tank. Furthermore, bathing was never a possibility on Mondays, which was the wash day for laundry. Mom switched that hot water tank off every evening before bed so all those who went to wash their hands would discover the faucet water was ice cold until noon the next day. So expensive were the school fees that George's mother wasn't even able to remain at home during the day, as other mothers on the block did. She worked as a bookkeeper for a local photographer and sometimes an African nanny took care of George and Marion after school, all because the older boys were boarding elsewhere, and Mom and Dad were working to pay for it. There was no doubt in George's mind: even if Dad did believe education to be the be-all-and-end-all, George thought the whole idea was just plain *dumb.*

Money troubles even seemed to plague them on vacation. To celebrate the previous Rhodes and Founders' weekend, the four-day bank holiday when Phineas typically left the Kelhams to travel home and visit his own family, and when it seemed like everyone in the whole world was taking a break from worries and indulging in fun, George's mother had simply packed a basket of sandwiches and the family had driven to Beira. They picnicked on the shores of the Indian Ocean, and then Marion and George swam in the large pool of the art deco-ed Grande Hotel, *The Estero,* the building's bare façade meant to showcase the Goliath as a symbol of unadulterated strength and success, so big and impenetrable it seemed almost like the place were a palace – though, one without a king.

The facility's grandiosity was all show. Technically, a person couldn't stay in the Grand Hotel in 1965 even if they wanted to. Its owners had priced rooms so high in hope of attracting the world's elite that regular families like the Kelhams – the folks most likely to holiday in Beira, as the city wasn't known as a hotspot for wealthier tourists – couldn't even begin to afford a night's stay. For two decades, the Grand Hotel had bled money without turning a single year's profit. But, instead of lowering the cost of its rooms to invite business from the typically middle-class families that visited the area, the hotel simply closed its doors in 1963. All that remained open was the Olympic-sized swimming pool over which the failed hotel imposed, casting shadow as it seemed to begrudgingly tolerate the locals who had never been welcome within its threshold. As George swam in the shade of that overbearing, empty building, it was almost as if it was simply one more person looking down on him, imposing restrictions, creating nonsensical rules and blankly sneering, *not good enough.* In that hotel, he saw every teacher who had ever cast a disapproving eye, every classmate who had mocked a hand-me-down shirt or secondhand book bag, and every time his own father had ever made him feel small as he towered above, judgement distorting his face into a mask of disgust. This thought made George angry, but it made him motivated, too. When young George looked up at that mighty building, he didn't know how he'd do it, but he did know he wouldn't let life always feel that way. Someday, somehow, George would find a way to rise above.

Dressed and assured of the day's good weather, George left his bedroom, entering immediately into the family's living room. He could hear that strange sniffling again coming from the kitchen, but this time there was a little cry, too. Suddenly, Phineas entered the living room from the kitchen, his head hung low, as though he were sorry for something. When the young man noticed George, he lifted his head in greeting, but his face was sad, his features flat. George's stomach turned. Something was definitely very wrong. Phineas was always smiling, polite and funny in his quiet way. Sometimes he even played with George, in a way that was actually work, so not to risk his position with the family. Once, George and Jimmy had collected a discarded baby carriage that they nailed boards too and rigged with a makeshift steering wheel, connecting the wheel to the tram's tires with rope. Phineas had helped the boys push the tram to the very top of the steepest hill they could find, so George and Jimmy could race their soapbox derby car and see how fast they could go. That time, Phineas had spent the greater part of the day helping push the tram up the hill, again and again, cheering the boys on as they rushed down.

Another time, George had gotten in trouble with Mom for getting ready too slowly in the morning.

"You are going to be late for school," she'd told him off. "You haven't yet eaten, and your laces aren't even tied! I want you out of this house in two minutes!"

George had sat paralyzed at the breakfast table, a piece of toast lifted halfway to his lips. Did Mom want him to keep eating, or to tie his shoes? Phineas had been the one to break the spell. Without saying a word, he'd ducked under the table and tied George's shoes while George, roused from his frozen indecision by Phineas' action, crammed the toast into his mouth. When George glanced down under the table to thank him, Phineas had just smiled back, and winked, and George had felt much better. For Phineas to be crestfallen meant that something was very, very wrong indeed, and suddenly, George felt very afraid.

George surveyed the scene through the kitchen door, which had been propped open. His parents and sister were sitting at the large wooden table Dad had built prior to constructing the back bedroom. The first thing George noticed was the strange look of his mother's face – her skin pale and blotchy, eyes red and cheeks wet with tears. She clutched a crumpled tissue in one hand. Mom wasn't sick, George realized, she had been crying. When she looked at George, her lips turned into a deeper frown and her shoulders started to shake, and, beside her, an awful whimper left Marion's lips, too.

"George," Dad said, his voice thin and whispery, like it hurt to speak too loudly. "Sit down."

George followed his father's order with slow steps, just like he took at church, because everything seemed very serious and not quite real, like the day required some kind of performance.

"What's wrong," he asked as he took his usual spot at the table. "Why is Mom crying?"

The way the muscles in Dad's jaw clenched made George's chest feel tight, painful with heartache. He was nearly horrified by the tears that began to fill Dad's eyes, not because they embarrassed the boy, but because it was such a terrible affront to see the strongest person he knew injured with such obvious severity.

"It's Francis, George," Dad said, and George's vision began to spin off-kilter just a little bit, the sight of the room around his father blurring at the edges as Dad spoke. The words coming out of his mouth weren't registering in George's head; it was as though they were muffled and nonsensical, like perhaps he was speaking a language George didn't know, or was talking under water.

"I don't understand," George said. "What happened?"

Dad repeated the awful tale a second time. The night before, Francis and the fellows he'd been celebrating with made their way to Mrs. Bolton's canteen in Umtali's main square. There, Francis had ordered a hot dog before tripping backwards and hitting his head on the pavement. Thinking it was nothing, Francis himself had stood up from the fall, shaking the dust off his jacket and laughing over his clumsiness with friends, but within the hour his speech had slurred, and it seemed he was having trouble walking. Soon after that his head had drooped, and it was as though Francis was having trouble staying awake. Members of the party made light of the situation, mocking Francis for his supposed drunkenness, calling him a lightweight. But Francis' closest friend of the bunch, the stag party's groom, knew Francis had only consumed two beers over the course of the entire evening.

Mrs. Bolton also knew Francis well. She knew the Kelhams and their children and had even babysat for John and Louise on occasion, when the family was younger. Mrs. Bolton insisted that Francis' friends take the young man to the emergency room, but they, who'd had quite a bit more to drink than Francis, insisted he was simply inebriated. Finally, a compromise was struck: the party arranged for Francis to be taken to the hospital outpatient center, where he could safely sleep off the drink, and he was admitted to the center as thus – the nurses believing him to be just another intoxicated young man on a Saturday night.

Francis had been found lifeless in an outpatient bed at seven-thirty the next morning. Upon medical inspection, coupled with Mrs. Bolton's eye-witness report, it was almost immediately obvious what had taken Francis' young life. When he'd tripped and hit his head on the street, a small rock had severed a main artery in Francis' temple and caused a major internal hemorrhage. Francis had bled to death in his sleep.

As Dad retold the story, Marion cried so hard that George had to raise his voice to be heard over hers.

"When do we go see him?" George asked. For a moment, he worried he'd said something wrong, and that maybe Dad was about to get cross with him. But then the tears that had welled behind his father's eyes overflowed and ran down his face, and Dad didn't even bother to wipe them away. For the first time in his life, George saw his father cry.

Mom reached over the table and took George's hand. It made perfect sense to George, why they weren't going to mass. They'd be going to the hospital to see poor Francis, instead.

"No, George," she said, her hand cold and shaky in his. "We won't see Francis again, son. He's gone to heaven now."

A lump started to form and hurt George's throat, and he wanted to cry too, but not because he was sad. George felt angry. He felt confused. He felt frustrated. Why weren't his parents making any sense? And why was Marion just *accepting* it? It was just like all the other stupid, nonsense rules: do this, don't do that, go here, stay there. Why were they not allowed to visit Francis? Why did Mom and Dad always have to be so strict?

"Why can't we just see him?" George demanded. "We could just go see him at the hospital."

He didn't know he was crying, too, until he heard his own voice, like it came from somewhere other than his body. The lump in his throat felt bigger and tighter.

"We can't, George," Mom wept. "We just can't. He's not there anymore."

George felt panicked, still trying to make sense of the impossible thing they were telling him when suddenly the front door opened, and George thought with devastating hope that it might be Francis coming inside, about to put right what had all been a terrible mistake. But no, it was Robert, who entered pale-faced and obviously shaken, the rims of his eyes also pink. Robert seemed to count the figures at the table, stopping when he came to George, his face falling.

"So, it's Francis," he said, as tears began to roll down his cheeks. "They just said it was my brother. I didn't know which one."

George got up from the table then, and pushed passed Robert, leaving the house out the front door. He walked and then ran until he arrived at Jimmy's house, and he barely had chance to knock before Mrs. Hynes opened the door to him and wrapped him up in her arms, pulling him into a tight hug.

"Francis –" George started to explain, but Mrs. Hynes didn't let him finish.

"We heard, son," she explained. "And we're so, so sorry."

George felt stunned when she released him. He looked around the kitchen he'd found himself in with almost no recollection of how he'd gotten there, all of Jimmy's family – Mr. Hynes, Angela, Jimmy, Brian – with so much pain on their faces, all for his hurt. George didn't know what to say, and Jimmy hung back, obviously unsure as to what to do. But Mr. Hynes moved in closer to George and his wife, and put his hand on George's shoulder, and bent down to George's level, and George was shocked to see tears in this grown man's eyes, too.

"A bloody awful thing, George. Oh lad, I'm so sorry."

The family welcomed George into their safe, warm home for the rest of the day, and soon Jimmy and George drifted away from the foreign state that was grief and into the familiar routine of play and exploration under the African sun. But every so often, suddenly and without warning, the terrible truth would descend, returning to the forefront of George's mind, and he'd remember Francis, and he'd remember that Francis was gone forever, and, inexplicably, George would think of the big back bedroom, which was now unarguably his and yet too large, haunted in its emptiness. And he thought of all of Francis' plans that were now all for naught, and the freedom and adulthood his big brother had just been on a cusp of enjoying, and George had a sinking feeling that perhaps all plans, and any idea of owning one's own life, was not as certain a thing as he had once imagined.

QUE QUE

1969

The drive to Que Que was long and miserably hot, and the Kelham family – then, only John, Louise, and George, as Marion had stayed over at a friend's home – passed the seven-hour one-way trip sticky with sweat. Mom sat in the front passenger seat, silent and seemingly exhausted. That was how she had been since Francis died: quiet, tired, and sad. It was the mood of the house and, apparently, it travelled with her.

Maybe that's the real reason why I'm being sent away, George thought. Maybe Mom is too tired for me now that Francis isn't around.

They were headed to Marist Brothers Secondary School, a place built one hour into the jungle, far, far away from George's home in Umtali. Once the vehicle reached this location, George would be left there, and he hated even the thought of it.

But the family wasn't about to travel that far in one day. About halfway through the drive, their vehicle stopped in front of a house George hadn't seen before. They were in Salisbury, Mom said, a city Mom and Dad had lived in for a brief time before Robert was born. The people inside the house were their friends and the Kelham family would be staying the night.

George was set up to sleep on a couch and, while the adults smoked, drank, and chatted around a kitchen table, George was placed in front of a black box set directly in front of the sofa. The box had a smooth screen that reflected at George like a dark mirror, and when the knob beside the screen was turned, the mirror flickered black and white. Once the flickering stopped, a man appeared in the box, sitting tall and handsome and well-groomed in greyscale, holding sheets of paper in his hands before him. George moved off the sofa and crawled close to the box, until his mother told him to sit further back or risk damaging his eyes. This was a television, George was told. The man was reading the news.

"It can't be good for a person," his mother said to the other adults as she walked back into the kitchen. George lost track of the adult's conversation in the background of his senses, his mother's voice turning into a mumbled murmur of words: radio waves, eye strain, radiation poisoning, cancer. George didn't care. He felt as though he had slipped into the pages of a sci-fi comic book. It was as though he had seen the future, and he knew it to be good.

But school, it turned out, was not at all good. If grief were a place, it was the Marist Brother's Secondary School: heavy, quiet, creepy, damn near inescapable. Its buildings were squat and faded in the usual African way, all colors stripped from coppery stone and painted boards by the unrelenting sun. It was run by a French-Canadian order of old Catholic priests who wore long brown frocks and the students had secretly nicknamed most of them – Smiley, Magoo – but none dared nickname Headmaster Charles.

While most of the pupils there were white upper crust, some were working class like George, and some were Black, Indian, and Asian – which was unusual in racially segregated Rhodesia – but few boys were ones George could call a friend. The Brothers didn't encourage friendship and they kept the boys so busy in schooling, worship, homework, and severely disciplined, mandatory sports – cricket, soccer, rugby – that there was no space for anything more than polite passing conversations between the students. Even hygiene, it seemed, was regulated. The Brothers insisted on supervising the boy's communal showers. Some of the students returned home to their parents every Friday afternoon, but George never returned home except for when school was out for its annual summer break.

In class, George and the other boys were to sit hour after hour in stiff desks, facing forward and keeping their mouths shut. When George failed to do so he received the strap, and soon after arriving at the school, his palms were made raw and tender.

Yet, George didn't feel he was particularly worse behaved than the other boys. His crimes were always petty: an untucked shirt, a scuffed shoe, a prematurely loosened tie. When other boys were guilty of the same, some of them were more likely to be reminded of the school's standards with a gentle hand laid on their shoulder, or a sharp look cast in their direction – no further redemption necessary. Those boys seemed to receive less homework than George and were sometimes given little gifts – candy and small playthings – from The Brothers. Some boys even moved around the halls after curfew, a rule George had never broken, and yet few ever participated in weekly detention as regularly as he did. If he had asked those boys why they crept the hallways at night, he knew they would have denied having done it at all, but there was no fooling George. He heard their sneaking footsteps near the passage that led to The Brothers' dormitory. It was easy to hear when all else was silent and he lay still and awake, tucked under his blankets.

For a long time, the only joy George experienced at Marist Brothers came when he'd return to his room to find a new letter from Jimmy Hynes. On those days, the exhaustion and loneliness would lift from George the way a dandelion seed parts from its stalk, rising to the sky, even if only temporarily. Those letters were evidence that someone in the world still knew of him, still wondered about him, and still wanted to hear from him. It was proof George was not bad, and that he still had a friend. No exhaustion was too great to keep him from answering those once-a-week letters. Regardless of how tight his shoulders might be set, or how sore his hands might be from punishment, George would pick up a pen and hurriedly reply to his friend's post. But then, about two years after the first letter had arrived, Jimmy wrote that his dad had gotten a new job and that his family was leaving Umtali. After that, Jimmy's letters failed to arrive. Outgoing mail was returned to George, Jimmy's new address unknown.

For a long time after, school was wholly joyless. Then one day, for reasons unknown, The Brothers organized a fair to be set up on school grounds. It may have been because parents with prospective students were visiting that weekend, or perhaps because an Archbishop or some senior member of the order was stopping by, but for whatever reason, the somber place was temporarily transformed into a land of play. There was a petting zoo and carnival games, and delicious fair food George could purchase for only a few cents. There was the opportunity to take a ride on a small, rickety Ferris wheel, too, which of course George took, but that was not the crown jewel of the event.

The star of the show had lowered to the ground like a gigantic, glittering insect, buzzing a terrible and exciting noise that filled George's ears and made it impossible to hear anything else. It was a helicopter, George knew, though he had never seen one in real life – a little two-seater which the pilot explained was known as a Bell 47. The boys were welcome to walk around the machine, but not to touch. George felt overwhelmed, as though there was too much to take in in far too little time. He wanted to press his nose up against each part of the machinery, to figure out how it all worked with the very same fascination he'd had when first viewing the television at his parents' friends' home.

"So, who's first to fly?" The pilot eventually said.

The boys around the pilot jumped up and down, their hands raising dollar bills in the air. George had no dollar bill to spend. He would not fly in that helicopter, but he would watch it lift and land, again and again.

But the fair was nothing but one rare carefree moment in three years of slog. Finally, the summer George turned fourteen, he refused to return.

In Rhodesia, summer break was observed over Christmas – the hottest portion of the year. Now in his mid-teens, George had aged out of the supervised, communal showers The Brothers' insisted on watching the younger boys take, but he still found the whole situation creepy and weird. George had also reached an age when he started to take notice of the adult world, and understand some of the adult conversation which, at earlier times in his life, had gone over his head. He knew about Brother Gaston and that the priest had been sent back to Canada because of a scandal involving the recent pregnancy of a school secretary. There had also been a news report about the Vatican and its priests' abuse of young men and boys. After hearing that, the showers, the footsteps in the hall, the discouragement of friendly conversation, and the teachers' habit of making pets of certain students suddenly made sense. Thinking on the subject, George was repulsed, infuriated, horrified, and sad.

"I'm not going back to Marist Brothers," he told his parents. He addressed them in the kitchen, standing with his shoulders pushed back, daring either one to doubt his resolve. George had worried about what his mother and father would say. Dad's feelings on education were obvious. He seemed to think it was nearly the only redeeming feature George might develop. For her part, Mom had sacrificed a lot for him to attend.

George was surprised, therefore, when Dad merely said, "That so?" barely lifting his eyes from the newspaper.

"Yes." George answered.

His mother sighed but didn't even break from cutting vegetables with Phineas in the kitchen.

"Well," she said, "I suppose we can register you at the local boy's school."

George nodded and was a bit flabbergasted, unsure of what to say. In hindsight, his parents may have already been considering bringing him closer to home. Throughout the nation, small pockets of violence were increasing in size. Perhaps his parents knew the chaos would only escalate.

Since Rhodesia's colonization in the late 1800's, those in governance had been white Europeans, who were a minority in the nation. This was largely due to voting restrictions; while all races technically had the right to vote, only landowners were allowed to cast ballots. However, much of the Black population lived on traditional tribal lands, therefore holding no private ownership. Further adding to the disparity, all races were allotted representative seats in the government, but a smaller amount were set aside for Black, Indian, and Asian representatives than were set aside for white Rhodesians. Had George been older and given full information on the politics of the time, he would have found the negative attitudes towards racial disparity difficult to argue, but George was very young as terror began to envelope his country, and young men – recruited for battle at eighteen years old, or even younger with parental permission – had neither power over political decisions nor the emotional surplus to invest into the issue as streets started to become littered with landmines.

Two separate armed groups had formed with the same goal, to eradicate white minority rule. They called themselves the ZIPRA and the ZANU, and white Rhodesia was beginning to hear of atrocities which, at the time George announced his reluctance to return to school, were predominantly inflicted on the Black populace the guerilla armies claimed to serve. A man and his wife were publicly executed because their daughter had married a police officer. Scores of schoolboys were abducted from their classrooms to be turned into child soldiers. Other parents were warned by the militant groups not to send their children to school – they considered formal institutions a shackle of colonialism and any parent that wished their child to receive such education a traitor to their cause. George hadn't paid attention to the changing politics of his country but, undoubtedly, his parents had.

That day in the kitchen, he had been primed to argue and was surprised no one wanted to fight. The work just kept on by the stove, and the reading continued in the living room. George went outside and took himself for a walk. He didn't know why his parents cared so little about his decision, but he didn't care, either. He had learned a valuable lesson.

One won't receive what one doesn't demand.

~

George's first day in Umtali Boys School began in January 1970. In many ways, attending high school in Umtali felt a little bit like George had gone back in time, back to the local elementary school with Jimmy. He was back in his hometown, able to sleep in his own bed and spend weekends swimming with friends. Freedom was within reach. Well, sort of.

George didn't make quick friends with the school's principal, Headmaster Coney Fleming. It seemed George was always in trouble for one thing or another – for poor grades, or poor attention in class, or even for being caught behaving recklessly in his spare time, uptown. Within only a few weeks of starting at Umtali Boys' School, blisters had developed into callouses on George's young hands from both punishment and work. Seemingly unable to rise to the level of obedience that Headmaster Fleming demanded of him, George spent a portion of every Saturday and Sunday in detention, planting trees along the perimeter of the large campus grounds or cutting back the wilder vegetation that crept in from the jungle from which the grounds had been carved. This was perhaps the only place where George saw himself represented at school. As aggressively as the school demanded the wild to yield, the wild simply remained, and steadily grew.

Still, despite the lengthy periods of free time George's detentions took up, the school was far preferred to Marist

Brothers. For as strict as the headmaster at the high school seemed to think he was, he was a kitten compared to those in leadership of George's previous school, and the sense of loneliness the former school's administration had fostered.

The Form Mistress at Umtali Boy's School, however, was another story. She *looked* like a kitten. Mrs. Monders was young, perhaps in her early or mid-twenties, with dark brunette hair, a figure that distracted from all her lessons, and seductive eyes. George near about fell in love with her as soon as he entered the classroom, but the feeling was short-lived.

Mrs. Monders' scorn was as sharp as her looks. It was clear from the flurry of hands that raised to answer her questions that Monders had fully infatuated her adolescent pupils, but she wasn't interested in meeting their crushes with kindness. She seemed to detest the boys, a fact which only served to make many of them only strive harder to meet her approval, but George was no stranger to disapproval from authority, and he wasn't interested in proving himself to anyone. Of course, Monders herself must have been unfamiliar to male ambivalence, as she met his disinterest with active, aggressive dislike.

She assigned mountains of homework every Friday, knowing George would struggle to complete the tasks and come back with only half done.

"That's typical you," she'd snarl at George in front of his classmates over his missed assignments, causing the other boys to laugh.

If George answered a question wrong, she'd laugh, inviting the other boys to laugh along. "Perhaps we can find a more intelligent answer?"

One day, as calmly and surely as though she were delivering a historical fact, Mrs. Monders looked George up and down and slowly shook her head.

"Mr. Kelham," she said, "you will never amount to anything. You'll be in jail before you're twenty-one years old."

Monders' blatant and obvious disapproval was shocking, but George wasn't completely innocent. He knew his sarcastic apathy drove her mad, as did the nickname he'd assigned her outside of class: Ma Monders.

There was one boy in particular who appreciated George's sense of humor. Neil Mayo, like George, had a sense of adventure and was always ready for a laugh. He had an older brother named

Clive and the boys were close enough in age that wherever one went the other was likely to follow. Neil and Clive were the only children of parents that were as equally friendly, intelligent, and fun-loving as the brothers were, and quickly Mr. and Mrs. Mayo seemed to adopt George as a third son, setting a place for him at the table to eat most dinners, happily expecting him to rejoin the family each morning.

When not hanging out at the Mayo's house or at school with Clive and Neil, the Mayo brothers and George would head downtown and meet up with Sean Griffith, another Umtali kid who had dropped out of school a few months prior having had locked down a good job. Now a working man, Sean was the only one in the gang who was truly free to behave however he pleased. He could grow his hair long, and never had to shine his shoes. The others – George, Neil, and Clive – had to remain constantly on the lookout for teachers or, even worse, universally-despised prefects like Jonathon Wince. If Jonathon caught you sneaking a cigarette or racing down main street after dark, all unruly behavior would be reported back to the headmaster. Fleming would then beat his wayward students with a cane, the severity of the beating in accordance with the severity of the boy's crime: a minimum of two whacks, a maximum of six. Fleming held his young students by the back of the neck and bent them over, whacking them on the ass as they yelped. Some boys would hide things down their pants to soften the blow: textbooks and the like. But Fleming would always check.

Then, he left bruises.

Jonathon was the eldest son of George's mother's employer, the town photographer for whom Louise kept books. Jonathon had a younger brother named Sebastian, a kid the same age as George but one George didn't know very well, but who seemed good for a laugh all the same. Jonathon was a different story altogether. Jonathon was like a mosquito in his ability to follow and annoy, seeming to have dedicated his life to patrolling the youth of Umtali in order to report petty crimes back to the headmasters. The kiss-ass deserved an ass-kicking, but George wasn't going to give it to him. That much of a sorry sad sack, it was bad enough just to *be* Jonathon Wince.

George's gang could usually be found hanging out at a small café called Meikles where they'd put whatever coins they had down on Brown Cow coke floats or, if they were really in the money, on the "world's best" waffles. When the boys had no more to spend, they'd exit the café but hang around the building, listening to music in the record store on the ground floor, or checking out sexy Mrs. Miller[1] as she did her shopping, or winking at girls their own age as they arrived for the tearoom. Those girls never gave them much thought. It seemed like all the girls in Umtali wanted a guy who could get them around – someone with a car or a motorcycle. And so, all guys in Umtali wanted a ride. Once, the guys happened to walk by a *beautiful* ride. George would later learn it was a Lamborghini Muira, but back then, all he knew was that it was gorgeous. It was also funny as there were long vents built in around the headlights that made George think of flirty, girlish eyelashes. George remarked to Neil and Clive that one day a car just like that would be his. His friends erupted in laughter.

Sometimes the guys would head to a restaurant and pub called The Wise Owl. No one cared much about the drinking age back then so Neil and Sean could usually score themselves a couple of beers, but George wasn't interested in that. When he thought of alcohol, two things came to mind: his father's miserable outlook on life and the way booze sustained it, and Francis, who, had it not been for alcohol and the assumptions people had made in his last moments, might have lived to be a world traveler and perhaps a husband and father, settled in a nearby home that George could escape to when Mom and Dad's was too stocked with sadness. Of course, home wouldn't be so sad, had Francis lived. No, George didn't touch anything stronger than the odd cigarette and a bottle of Coca Cola.

One day, while the boys played pinball at a café called Pop's, a guy in his mid-twenties headed over to the table to say hello to Sean. Sean introduced the man as Vic Jenkins and said he was a linesman at the electricity supply commission when Sean worked.

"You hiring?" George joked.

Vic didn't see the humor. He nodded his head.

"Always," he said. "You lookin' for work?"

George didn't know what to say. He hadn't really considered that he'd be welcome to start applying for work at only fifteen years old, but obviously Sean had already chosen to leave school, and while his parents wanted him to graduate, it was obvious

[1] Name fictionalized.

George wasn't on route to college. He'd started out at Umtali High School in the academically focused Form 3A with Neil but had dropped a level in scholastic performance almost every term. Now that George was in a program for students who were most likely to find a future career in the skilled trades, he could no longer see the value in attendance. The trades didn't require a high school diploma. What was the point of wasting time to get one if there was a job being offered right now?

"Yeah," George said. "I am."

"You are?" Neil asked.

"Right on," Sean said. "You'll love it, George. Great money."

"Got a pen?" asked Vic. Neil handed him one and Vic bent down over the table, scratching digits onto a paper napkin. "Tony Cappa," Vic said. "You call his number and tell him I sent you."

George didn't waste much time before going home and dialing Tony's number. He stood just outside the kitchen, in hope and disbelief as the line rang, the curled phone line stretched around the corner from the receiver to where George stood as far away from his mother as possible for privacy. He scratched Tony's office address on his hand in blue ink when he was invited to interview the next week.

Unbelievable, George thought, hanging up the phone. How was it so easy? How didn't he think to ask Sean to help him land a job earlier?

Later that night, George told his parents he had secured an interview. Somehow, he convinced his dad to take him and while John Leonard was obviously displeased with his son's choice to drop out of school, he didn't fault George's logic.

"Wear a tie," he told his son when George emerged from the back bedroom in slacks and a shirt. "It makes an impression."

George was surprised by the gesture of support. He nodded. "Okay. Thanks."

They arrived in Salisbury that afternoon. John Leonard came into the meeting with George and introduced himself to Tony, with whom he shook hands before excusing himself from the room, leaving George feeling a little unprepared and shocked. In George's eyes, Tony Cappa was the kingpin of the electricity supply commission. George stumbled through the answers to Tony's interview questions as best he could, hoping he was saying the right things. Finally, the questions stopped coming. George had nothing else to say, so he sat there, waiting for Tony to pass judgement.

Tony looked at George for a moment and then shook his head and sighed. He stood up and extended his hand over the desk.

"Well, Mr. Kelham," he said, "you are a little young, but we'll take you anyway."

George walked to his father's car in a state of total euphoria. With the shake of a hand, he had transformed from a schoolboy to a working man.

He opened the passenger side door and got in. John Leonard didn't say anything. He just put the gear shift into drive.

"I got the job," George finally said.

"Well, that's that, then." John Leonard said in a low, disapproving voice. "You've made your decision."

They drove home in silence.

~

George was fifteen years and ten months old when he started work on November 1st, 1970. His mother, unconditionally proud of and happy for him, helped George pack his bag.

The Electricity Supply Company (or more commonly referred to as ESC) was putting power lines throughout Rhodesia. George's job was to do what he was told and go wherever they placed him – usually in the bush in the middle of nowhere. He worked Monday to Friday, returning home when he was able to catch a ride, or staying in camp for the weekend if nobody was heading into town. While there, George stayed in a camper owned by the company. The men worked in crews of fifteen and did everything through manual labor. On each crew, there were teams of white workers and Shona laborers. The laborers dug the holes with shovels and hoes and the workers lifted and lowered the heavy poles into the holes, filling them back in with the earth that had been moved. The next week the team would run the conductor, linking the power poles with line, and then, the week after that, the process would start all over again. It was hard, physical work in the hot, dry sun and even despite George's youth, by evening he was ready to collapse onto his thin camper mattress, every muscle in his sore body crying out for the relief of sleep.

But it only took a few weeks for George's body to develop the muscles needed for the job, and then he acclimatized to the work, staying up later in the evening and getting to know the other men in his crew. His supervisor was a man named Chris Johnson. Johnson was a rugged, friendly man who brought his whole family to camp with him. He had three daughters who played together around the camp site, and the eldest was very pretty but George was crippled with shyness whenever she said hello. Mrs. Johnson was warm and welcoming and an excellent cook. The men were each paid one hundred and twenty dollars every month, plus an extra dollar for every day they spent in camp.

George and every other man in camp gave their daily dollars to Mrs. Johnson and in return she kept them well fed with heaping plates of steak and potatoes, burgers and fries, eggs and sizzling bacon. After dinner, the men would play cards; friendly, pass-the-time games without money put on an outcome. At camp and away from most opportunities to spend cash, George quickly saved enough to buy a motorbike – a gorgeous gold Honda 750. After that, he spent his evenings riding the dirt roads all around the camp before returning to have a few laughs with the other men around a hand or two of poker. On Friday nights, he was able to travel home to Umtali. He grew his hair out long and no one cared if his shirt had become untucked or if his shoes were scuffed. It didn't matter what he looked like in the bush. Sweat poured down his head and stayed there in dark streaks until he had time at the end of a long, hard, satisfying day of work to wash himself clean. Camp life was the exact opposite of his life at Marist Brothers and Umtali Boys' School; it was all laughter, all action, all freedom. George was the happiest he'd ever been.

At home on days off, George embraced the counterculture fashion of the 1960's, happy to have replaced his schoolboy ties with flower-print shirts. He had money enough to buy dinner at the Wise Owl restaurant whenever he liked and snorted with bitter laughter when Neil told him Mrs. Monders had simply replied 'good riddance' when Neil announced that George had officially left school. George listened to Led Zeppelin and The Rolling Stones and would have worn bell-bottomed jeans and corduroy slacks had the country not been so bloody hot all the time. George, Sean, and Neil rode for hours on their motorbikes each day, and once he saved enough money, George bought a little car: a secondhand beach buggy that was always in need of repair. The gang would park outside the Dairy Den ice cream shop and then when they winked at girls, the girls started smiling back.

Soon, George met Hannah Leary[2].

Hannah was a beautiful girl with brown hair and a smile so big her eyes all but closed when she was truly happy. She was the kind of girl who was often truly happy, and who looked for good in the world and for the good in people, and, as pretty as she was, she was made prettier still by her kindness. She seemed to think George was a good person and she told him that he seemed older than other boys their age, more mature. George thought the same of Hannah. Despite her aptitude to cheerfulness, she was serious in a way other girls weren't, never silly or petty. George and Hannah were kind and good to one another.

It was a puppy-love romance at first. George asked her to go to a movie, and immediately they were an item. The young couple spent most of their time together at friend's houses, or swimming in local swimming pools or hot springs with a group of teenage friends. Hannah's parents were concerned about their age; George was sixteen and she was fifteen, but the Leary's tolerated the relationship well – exceptionally well even, considering the class difference between them and the Kelhams. Mr. Leary was a Police Inspector for the city of Umtali and was a stern man, loving but strict. Mrs. Leary, however – possibly because she was every bit as kind as her daughter – went out of her way to form a casual friendship with George's mother, stopping by Rudland Avenue for tea every now and then.

On special occasions, the lovebirds gave each other gifts: Hannah knit George a sweater that matched her own, and he bought her a huge stuffed dog that was red and white in color. One day, when George's unreliable beach buggy was actually running, he surprised her by driving her to Rusape, where her younger sister was buried. The pair bonded over that: each of them bore the weight placed on children upon the loss of a sibling. When George and Hannah were together, grief wasn't hard or silent or sick. It was as familiar and benign as a head cold, something they could nurse one another through, and recover from, healthier on the other side.

Largely, the couple seemed to live in a paradise of their own making. But Umtali wasn't paradise, and neither was Rhodesia. Things were unravelling in ways that George hadn't paid much attention to before, but now seemed impossible to ignore. It

[2] Name fictionalized.

started with the news that would broadcast through his mother's kitchen countertop radio. Reports of linesmen, like himself, ambushed by violent groups some nations called freedom fighters, and some called terrorists. Rural and tribal lands were being littered with landmines and he sometimes overheard adults speaking of massacres: farmers and their children brutally killed before their homes were ransacked for valuables: money, food, cattle, crops. Many farm laborers – most of them Black Africans – were murdered with gratuitous violence, their bodies strung on road-facing fences as a message: any association with colonizers was considered treason and worthy of such atrocity by liberating forces.

George wasn't interested in politics, but he knew Rhodesia had declared a unilateral declaration of independence from Britain since the Queen's country had demanded all African nations grant every citizen full suffrage. Ian Smith's government and many Rhodesian citizens, fearing the violent chaos that the Congo had experienced soon after moving to satisfy Britain's demands, voted to withhold full voting rights and, because of that, the country was considered an illegitimate state under strict trade sanctions from America, Canada, and most of Europe. Cigarette lighters were impossible to find, so people used matches. New vehicles could only be purchased out of country. The government started rationing petrol, so George and Hannah would take twice-monthly trips over the border to Mozambique where he'd fill his gas tank and several jerry cans to take home.

As these reminders came up again and again in the background conversations of George's life, he was constantly, rudely, reminded that he was not yet, in fact, truly a free man. There was still the issue of military service and if George wanted to keep living in Rhodesia, there was no way out of it. Francis had completed the four and a half months of training that was required of young men during his time. But since that time, service had ramped up in relation to increased insurgent attacks and now service was a mandatory four and a half months of training, plus an equal tour of duty – a responsibility Robert had recently undertaken, and George had yet to fulfill.

Nine and a half months seemed like an impossible duration at George's young age. The idea of it buzzed in George's mind like a fly at a picnic, spoiling what should have been such good fun. Until his service was completed, it was unlikely George would be promoted at work. ESC wouldn't want to invest in someone who was legally bound to leave them. But afterwards… well, servicemen were awarded greater respect. George might even have the opportunity to become foreman of his own crew! And if he were promoted, well, then he'd really be making money.

Conscription was just something that had to be completed. Something that had to be done and over with, George finally decided, speaking on the subject with his friends. Clive and Neil were still in school and therefore exempt until after graduation, but service was an obligation that similarly bothered Sean. Together, George and Sean decided the ordeal would be made easier by the comfort of one another's company.

The only problem was George's age. Sean was a year older and able to enlist on his own volition, but although George had been working fulltime for more than a year, he had only just turned seventeen years old, and would therefore require his parent's signed permission.

It was easy to obtain the paperwork indicating parental permission for George to enlist as a minor but having them sign it was a different story all together. He knew the forms required the signature of only one parent and that his mother would support whatever choice George made, but he also knew his father wouldn't approve, and that it would be nearly impossible to catch Mom alone and away from John Leonard as the Kelham home was so small. George was considering how he might orchestrate the thing one Saturday morning when heading into Umtali on his days off from ESC bush camp when he happened to run into his mother while he had the papers folded in his back pocket.

Louise was on the sidewalk outside of the bank. George merely turned the corner from a retail store and there she was: by herself and happy to have unexpectedly bumped into her youngest son. Louise Susannah hugged George hello. He showed her the paperwork, she signed, and then it was all done.

George mailed the papers away immediately, finally able to put the problem of when to enlist out of his mind. Like all boys his age, the future still seemed like a place far, far away, but, within only a few weeks, Louise called George at work to let him know his call to conscribe had arrived in the post. The military had sent George his first orders: it was time to enlist.

When George returned home to answer the call his father was

sitting in his chair in the living room. George said hello and explained why he was home, and then left to the back bedroom to unpack his things.

"They don't call boys up until they're eighteen," George overheard his father say to his mother, but George pretended to ignore where the conversation was heading.

"How was he let in without our permission?"

But it didn't matter. What was done was done.

~

George was given two weeks to get his life in order before leaving for basic training. First, he made arrangements with his employer. Tony was not happy to hear George was going into basic training as George was a good, hard worker who would be sorely missed in camp, but they valued him enough to want him back, afterwards. To that end, they promised to continue to top-up his military wage so he wouldn't lose his regular income while in duty, so long as he agreed to return to fulltime work after his training and subsequent tour was complete.

The next thing George did was ensure his mother had access to his bank account and necessary information: his passport, his bank book, tax forms. He told Louise that Hannah must be able to access the money and information, too. He'd be relying on his girlfriend to forward whatever funds he found himself in need of, and it was important to George that if Hannah found herself in need, too, that money was there for her to use. If worse came to worst, George would not have what little he'd earned languishing in an inaccessible account. That money could be split between the only two women who had ever loved him.

Hannah, in those two weeks, wept at the drop of a hat. She hated the thought of George going away and tied too much meaning to each little thing they did.

"This will likely be the last time we get ice cream before you go away," she'd say, when visiting the Dairy Den. If he took her to the tea house, instead, she'd say, "Basic training might be over before we're back here, again."

George reminded her every time that he'd be gone only for training and one mere tour of duty, but the thought provided little console.

"You'll forget about me," she cried one night, sitting in the old beach buggy.

"I won't," he promised. He was sure he wouldn't, but still the promise was made with some amount of boyish bewilderment. George was unaccustomed to such big feelings. He loved Hannah, but George's mind was more consumed by the idea of training camp and war than of romance. Whatever future they had seemed so far away it was impossible to imagine, never mind predict. But as she cried into a tissue, George had an idea.

The next day, he instructed Hannah to meet him at Mr. Wince's photo studio.

"What are we doing here?" She laughed, to be brought to such a random location.

"Hiya, Love," Mr. Wince called as he came out to the front of shop.

"What do you think?" George asked. "It's a photography studio, isn't it? I need a photo of my lady to take to camp."

Shy, pretty Hannah sat in front of a backdrop as George stood a few feet back from the scene. Mr. Wince was generous, taking photo after photo, arranging Hannah in becoming and demure poses. She was delighted to have such attention placed on her and laughed and smiled and blushed at Mr. Wince. But every so often, Hannah looked towards George, and then a different sort of smile flashed over her face. When it came time to pick which portrait George would have developed, George chose that one, the girl's eyes drifted away from the photographer and camera to someone just out of frame.

The day before George boarded the train to military camp, Hannah's parents gave her special permission to share Marion's room overnight at the Kelham house, so that the young couple needn't spend their last night apart – though, technically, they were in different rooms. That day, Hannah arrived early in the morning, and helped Mrs. Kelham and Phineas get breakfast on the table.

After eating, George gave her a peck on the cheek and told her he'd be back shortly.

"Where are you going?" She asked, surprised.

"I'm meeting Sean to get our haircuts," he said.

"Oh!" Hannah said, as though she'd forgotten that was a thing – the last thing – he still needed to do. "Do you want me to come with you?"

George shook his head and laughed. "Gotta do this one alone,

I'm afraid."

Hannah nodded, but ran a hand through his long hair, pushing it back from his face.

"Gosh," she said. "It sure will be a change."

Sean was already sitting on the curb beside his motorbike outside the barbershop when George arrived on his own machine. Sean looked downright miserable, smoking a cigarette and dangling a beer bottle between his bare knees as he waited in the hot sun.

"Ready?" George asked, taking the cig from his friend for a quick puff before throwing it away.

"As I'll ever be," Sean darkly replied.

The boys headed inside the barber shop. As George sat under the buzz of the barber's shears, he imagined the whirling was something else: a bike, a buggy motor - anything other than this little tool stripping him of the identity he'd created, his long hair proof of the first choice he'd ever been able to make for himself. He stared into his own blue eyes in the mirror as the hair fell from his head, onto his shoulder, and finally to the floor. When the barber was done, George had been turned into a person he didn't recognize.

The boys stopped by the ice cream shop before George returned home for his last evening with Hannah. George and Sean hung their freshly bald heads as they licked at their cones, unnerved by the country so determined to make them into men.

LLEWELLYN BARRACKS

February 21st, 1972

The next morning, George, Hannah, and Louise Susannah drove to meet Sean Griffith at the Umtali train station. There were hugs all around, and the women cried, and Louise held George in a tight hug before he boarded the train.

"I'll pray for you," she whispered in his ear. "I'll never stop praying for you, Love."

The boys wore plain clothes but were marked for military service by their crewcut hairstyles and age. They boarded the train with civilians but were directed to a military carriage with other Rhodesian boys on route to basic training. Once seated, the carriage was quiet and still. The boys looked out the window and watched the women they loved – the women who loved them – become smaller and smaller, their faces turning unrecognizable in the distance but their bodies never turning from view of the train, not until they were out of sight. Still, George imagined, *still* his mother might be standing on the platform, holding vigil for the life of her youngest child.

I'll pray for you, she had said. *I'll never stop praying.*

Despite his years as an altar boy and his dramatic immersion into religious schools, George didn't believe in God. Not really. But he did believe in his mother, and in her love for him. And by it, George felt safe. Protected. Like someone was watching over him.

The journey was long. Sean and George sat opposite one

another but had little to say. They stared out the window and groaned at every stop, each extra small, rural station adding hours to the trip. In all, they travelled for two days and one night. On the second day, knowing that would be the day they arrived at the basic training camp, George retrieved Hannah's photograph from his bag and placed it at the bottom of his shoe, between his socks and its sole, hoping to keep the image on his person. He'd been warned by his older brother, Robert, that new servicemen were stripped of all belongings. In training, a person wasn't a person. There was no room for a cross pendant, or a letter from one's mother, or even the memory of a girlfriend. In basic training, a serviceman was no more than a number.

"Just keep quiet and stay in the middle of the pack," Robert had advised. They'd been hanging out in the back garden when Robert was telling George of his own experiences in training camp. "Right in the middle is where you want to be. The guy in the front, he gets extra work. The guy in the back, he gets shit. Just fade into the middle. Don't stand out."

Finally, the train stopped in late afternoon. Outside the window, the landscape was reduced to dark jungle, the world only black and green, shadows made by leaves as large as George's torso. The civilian carriages were empty, George noticed, as he slung his pack over his shoulder. He followed his friend off the train into the hot sun, legs aching from sitting so long.

They had seemed to have been let off in the middle of nowhere. There was no platform to step onto, only grass, but a line of military vehicles was parked opposite the train in a ditch. The boys were instructed by a man who shouted out that all soldiers were to find a seat in a vehicle.

George and Sean found a bench seat to share in the back of a tarp-covered truck and tried to remain as straight and motionless as possible as the vehicle rocked down a dirt road. Despite the truck's canvas top, the sun was bright enough to light up the space where George and Sean sat. Discreetly, George scanned the faces of other young men for any clue as to where they were going. Their faces seemed pinched, anxious. Suddenly, George remembered the war was all around them.

Are we supposed to keep our heads low? George wondered. Might we be shot at, right now?

After a long, tense drive, the canvas top alternating between flashes of sunlight and the shadows of tree limbs, the vehicle finally stopped. Outside the canvas top, George could hear other car engines ceasing and the voices of loud men. It was clear they had arrived at the training base before orders to exit the vehicle were even barked.

But barked, orders were.

"Out, out!" Voices roared at the boys. George and Sean shuffled out of the vehicle as fast as they could, George swinging his pack over his shoulder with such gusto the weight of it nearly toppled his five-foot seven frame.

He found himself in a gravel lot amidst a hive of activity. Most service men wore uniforms and moved with quick intent, seeming to know exactly what task they were expected to perform at any given time. In his peripheral vision, George saw other young men his age running in tandem around the perimeter of the vicinity. George felt completely green, standing wide-eyed and unsure in plain clothes beside Sean.

"Line up!" A commander screamed as George and Sean scrambled to meet his demands. They hustled before the senior official in total confusion. Where did the line start? Were they meant to establish one?

Neither the front nor the back, thought George, remembering Robert's advice. George hung back, letting other recruits make up the first line. He stood straight with his eyes forward, his chest out, avoiding all eye contact with the screaming commander.

They knew they had been assigned to C Company, 123 Intake from the papers they'd received in the post, prior. Before long, George and Sean were given their uniforms and shown to their barracks. In the barracks, the boys were told to hand over all the clothing they had worn on their person when they arrived, as well as their packed belongings. The items would be inspected for contraband, and the clothing would be cast aside. Only what the military decided was truly vital for a person's survival might be returned.

George scanned the barracks quickly as he changed into his uniform. There were eight other young men in the dormitory, but no time to chat.

Once dressed in uniform, with Hannah's photo pushed again into the bottom of his boot, George left the barracks to line up with other young men in the parade square.

"You are scum!" The commander screamed at the boys, spittle flying from his grimaced mouth. "You are the worst group of weak, cowardly, shit-eaters this base has ever had the misfortune of training. Am I right?"

"Sir, yes, sir!" The crowd of boys responded, George's voice raising to meet the rest. His eyes kept straight in fear of being noticed, he wondered what Sean, standing next to him, thought. Had they been home in Umtali, and the commander a teacher instead of a military superior, they likely would have burst out laughing at the madness of this scene, at the feigned outrage of the commander's tone. George tried to put the thought away lest it force him to smirk without realizing it. He did *not* want the commander to catch him smirking.

Dear god, he thought, *do not let me be asked what I find so funny.*

By the time the commander had finished berating the new troops, George felt as though he might have sweated through his new uniform. Beads of sweat rolled down his face under his hat, and the hairs on his bare legs prickled as the sweat caught mild breezes that blew in too softly for relief.

"Turn!" The commander screamed, and for a moment, George wasn't sure what the word meant. He lost his straight-forward focus to look at the other soldiers and saw those to his left pivot so they faced left, and so George did too. Then, one by one, each line began to march, so George mimicked that, too, copying the person in front of him, hoping that the recruit he followed knew what he was doing, hoping that Sean, now behind George, wasn't copying him as closely as George was copying the boy ahead.

The new troop was led around the perimeter of the base, again, and again, and again. Finally, when the sun was low in the sky and the uniform that had first seemed too bulky to wear in the great summer heat was now feeling too little, too sweat-soaked, and too thin to prevent hypothermia, the troop was dismissed to find their supper at the mess hall. After that, they were ordered to shower and then to polish their shoes. Finally, at midnight, the young men were generously bestowed the privilege of sleep.

~

George had known hard labor. His first weeks of employment, after all, had tried his body in ways he had never been tried before; breaking him down before rebuilding him, stronger. But it was not until basic training that George had known complete and total exhaustion.

In basic training, all recruits were as one. They were all young men or, like George, still teenagers, yet the burgeoning individuality typical of early adulthood would not be humored. Names were unnecessary. George no longer existed. Sean no longer existed. Some of the recruits had come voluntarily from other nations – South Africa, Portugal – and others were American vets still fresh from Vietnam, unwilling or perhaps unable to put war behind them. None of this mattered. The men were only Company Intake C; their lives reduced to each fleeting moment, to each immense task, their state-owned bodies no longer human but seemingly thought to be more machine than man by those in command. Recruits woke at the same time, showered at the same time, ate at the same time, shit at the same time. They woke before the sun and marched until steam rose from their hot skin in still-frigid morning air. Beds were identical packages, sheets and blankets measured for millimetric accuracy by tight strings held from the first bed to the last. In mid-morning, they cleaned all areas of the base from top to bottom: their barracks, the mess hall, the showers.

After a few weeks, the men were instructed and tested on their firing skills. Simultaneously, they fired rifles, machine guns, pistols, and practiced throwing grenades. They practiced taking their guns apart and putting them back together. They reconstructed machine guns while blindfolded to ensure each soldier could repair it in the dark if need be. On one testing day, George lined up alongside sixteen other recruits on the range and shot machine gun rounds into the targets before them, as best they could. The machine in his hands seemed impossibly heavy. When he held it parallel to the ground, his arms shook with the effort of raising it up and nearly crumpled while holding the trigger during its rapid release. He shuddered at the thought of having to pack such a weight through jungle terrain but took solace in the fact that he had no talent or experience in handling weapons, and therefore was unlikely to be given the responsibility of such a significant item.

In afternoon, the recruits marched again until the high sun exhausted them, and then they would stand for hours at attention, expected to digest information George would only later understand as propaganda. The enemy was supported by communists, the soldiers were told. This was what the rest of the world failed to understand. The world stood on the brink of a cold war and the terrorists were funded, outfitted, and trained by international threats: Russia, China. But their weapons were inferior. Insufficient. Rhodesia was winning the war, and there was no doubt it would be won.

Then, brains full and bodies nearly crippled, the boys would train and march and clean, again.

One of the training drills the recruits routinely practiced did pique George's interest. For hours at a time, the troop modelled how to safely board and exit the body of a helicopter, accounting for the fact they'd likely be shot at while exiting, or quite literally lose their heads should they jump out with too much excitement. The training helicopter was only a grounded hunk of steel, set aside for drill purposes only. The retired machine had been stripped of rotor blades and engine, and the practice itself was a bore, but George was intrigued by the idea of how such a heavy machine could freely lift and propel upwards, into the sky. He thought of the shining craft that had visited Marist Brothers, and how he had spent the day watching richer children embark on an adventure that had been withheld from him. He imagined the machine taking him away from the base, from the planet, as though he were weightless. He imagined that was what true freedom must feel like. If the cost to fly was to risk being shot at, it was a price George was willing to pay.

In those first few weeks, George seemed to disassociate from himself. He knew it was some act of physical self-preservation that his body seemed to shut down all functions necessary to enjoyment, but because it coincided with the removal of his identity by military command, the experience had an eerie affect. George's mouth stopped tasting food. His stomach, though empty and surely calorically depleted, barely called for sustenance at all. The lukewarm water that passed over him in the communal showers was not felt on his numb skin. When C Company Intake 123 was finally bestowed the privilege of sleep, George's nights were dreamless.

But this was the point of basic training: to make machines from those who had once been men. This would be George's life for four months and two weeks.

Then, just as his body had recalibrated to meet the work of laying electrical lines, George eventually found exhaustion slipping away. He and the rest of the recruits soon found they needed less sleep than assigned, and soon, minute by minute, evenings grew past the daily task of shining shoes. Those late hours, stripped of the comradery, comforts, and lightheartedness George had known at ESC, were as though a shadowed version of camp: they were simply time that demanded to be filled. Some men wrote letters home. Others played quiet games of cards. George, as he had done in the hours after building powerlines, found a way to fly.

He took to the perimeter of the base, the same route he marched every day, as the boundary was as close to freedom as he could get. Lacking the motorbike that had helped him fill evening hours when working for ESC, George called upon his own body for speed. Although each day tired his legs, and what little weight he'd arrived with on his 143lb teenage frame had already melted away to malnourishment, George ran. He ran in the setting sun, the air cooling, catching the sweat of his skin. He ran alone and ran to feel the speed in stark contrast to how slowly the hours of his life then seemed to pass. He ran to feel his legs burn; he ran to know his body simply by proof of its pain; he ran for evidence that he was more than a number. He ran to leave behind the weight of obligation, duty, and law, that bound him to the earth – to this *specific* piece of earth – and to a war that wasn't his.

In this way, George flew.

Finally, after four and a half months, the day had come. The recruits stood at attention in the parade square and were led through a ceremony that signified their change in status from scum to soldiers. Afterwards, George and Sean learned that they were both to be transferred to the same base, though in different outfits. Sean, with a knack for communications, would serve in signals and George was to be on foot patrol, the bottom of the barrel in military rank and choice of task.

The soldiers were given marginally more rights, as well as the opportunity to return home for a break before beginning active duty. George was presented with a stack of letters from Hannah, one for almost every night he'd been away. While he had ran, she had written. But it was faster to go home than to write back.

George hitched a ride with a soldier from a neighboring barrack who had driven his Volkswagen Beetle to base instead of taking the train, as most others had done. For one dollar, the man took George from Bullawo to Umtali with five other homesick young men crammed into a backseat that was meant to serve three.

"Coming?" George asked Sean as he counted out enough change to pay the driver.

Sean shook his head. "I just want to sleep, man. Just two days in my cot, closing my eyes whenever I want."

Hannah was waiting at the Kelham house when George arrived. She chastely kissed him hello and then moved aside for Mrs. Kelham to step forward and take her son's face in her hands, inspecting his eyes for wellness, for a glimpse of the child she had known. But George pulled away. He wasn't a child anymore. He was seventeen, and a soldier. He'd lived by himself for two years by that point, if not in the barracks, then in the ESC camp.

Later that night, George and Hannah went to a drive-in movie and then met the next morning for breakfast at the tea shop before George caught up with Clive and Neil for cokes at the Wise Owl. Before George knew it, it was Sunday morning. The soldier that drove him home returned for George at their meeting place, and he and five other young men made the six-hour trip back.

KARIBA
July 1972

The training base transported new soldiers by military vehicles to their new bases for their tour of duty. George and Sean were taken to Kariba. Already dressed in uniform and seasoned in military protocol, George arrived on Kariba's parade square with some idea on where to stand and how to behave, a stark change from arrival in the first camp.

After receiving a barking introduction like the one he had received on his first day, the soldiers were excused to their bunks, and George tucked his bag under his new bed and slipped Hannah's photo under his pillow. Next, he reported to the armory to collect the firearm he would be charged with keeping on his person until his duty was complete.

He waited at attention until a man shouted his name.

"Kelham," a man barked.

George squinted against the sun and stepped forward to collect his rifle.

"Sir, yes, sir!" He raised his finger to his forehead in salute.

"At ease, marksman," his lieutenant replied.

George fought the urge to correct the lieutenant. George knew he was a rifleman, and most likely second class, at that. But something was wrong. George watched in unexpressed distress as the arms man stepped up beside the lieutenant, brandishing a machine gun.

"Congratulations, Mr. Kelham," George's superior barked.

"You have been examined and found to be an expert marksman, soldier. You have the privilege of carrying on your person this powerful piece of military artillery and all complimentary ammunition. This is an honor and an order, soldier."

George's heart was beating out of his chest. The machine gun itself weighed almost twenty-eight pounds, and the ammunition would add another ten. George himself weighed only one hundred and forty-three. How would he carry the damn thing through rough terrain? He wasn't a marksman. Something was wrong.

"Sir, yes, sir!" George heard himself exclaim. No other answer existed.

The artilleryman passed the machine gun to George and George felt his shoulder ache immediately upon slinging the thick strap. Then the artillery man passed George a backpack stuffed so full its contents seemed liable to split the canvas' seems. George did his best to refrain from grunting as he lifted the sack, full of metal ammunition.

The rest of George's day was spent performing drills and memorizing the routes he'd be patrolling. Each tour of duty would have him rotate through one of three locations: the flatlands along the Zambezi River, the wall of the hydro dam within the city of Kariba, and *the gomos* – the thick jungle beneath a mountain top radio tower. The new servicemen would keep these routines for four and a half months before returning to their hometowns.

After a miserable, silent supper, George returned to the barracks, dropping the machine gun and ammunition at the foot of his bed. He removed his uniform shirt and tried to loosen the muscles in his upper back, rubbing his arms and shoulders with his hands.

"Heavy, huh?" One of the soldiers asked.

"Yeah," George replied. He was tired. "I don't know how I got stuck with this bitch. I know I'm not a good shot. I should be a rifleman." In each troop, only the expert marksman would carry the machine gun and ammunition. All other soldiers would be responsible for their own lighter rifles, and for taking turns carrying the heavy radio pack.

On the other side of the barracks, a group of soldiers laughed among themselves. Every so often, one or two would look in George's direction, and George got the sense he was the butt of the joke.

"Didn't anyone back home warn you about the range exam?"

the soldier he'd been speaking with asked.

George shook his head. The soldier lowered his voice.

"Who were standing by during your shooting tests? Them over there?" He jutted his chin in the direction of the others.

George was confused. He felt his brows knit together with almost the same pressure he felt in every sore limb.

"Yeah," he shrugged. "Yeah, I was between them. How'd you know?"

The man shook his head. "George, they were shooting at your target. No one shoots at their own target. Not if they know better. Who wants to end up lugging that thing around the jungle?" He nodded to the automatic and its ammunition.

George was choked with disappointment. He hadn't known and now he felt a fool. He could have split someone's lip for it, but that would only invite a heap more trouble than he was already in. Exhausted and disappointed, George got into bed and stared at the ceiling, willing his muscles to ease at least enough to achieve sleep.

The next day, George was issued a rat pack: a week's worth of dehydrated food so tasteless the small canteen might as well have been filled with biscuits of drywall. George unpacked the food to repack the sustenance more efficiently. He had packed light, knowing his burden would include the machine gun and its ammunition. Still, as much as his own body weight was strapped to him in dehydrated food and water supplies, a machine gun and bullets, spare socks, and a sleeping bag.

"You may pack your damn beds if you can carry them, servicemen," his superior barked as he stood at attention. "But anything you take you must carry along with the responsibility of your assigned weapon. You will not return to base for at least one week. I suggest you choose wisely."

George knew his rotation would begin in the flatlands. To get there, the troop was taken by military vehicle to a helicopter pad on the outskirt of the base.

And there they were, sitting on the pad with the grace of beauty queens on stage: Alouette IIIs, George learned they were called. Light utility helicopters designed for quick and nimble military transport, their steel blades shining bright in the sun. These were nothing like the bare-bone steel frame he'd practiced getting in and out of. The Allottees were sleek and strong, robust while still light. As the soldiers arrived, their blades slowly started rotating, as though the machines were happy to see them in the same way a dog might wag its tail upon its owner's return. Those blades were hypnotizing, speeding up as George exited the jeep.

Suddenly, a hand slapped his back.

"Go, go, go!" His commander shouted, bringing George back to reality. Each chopper could fit four soldiers and, as the machinist, George was the last to board in order to sit by the open door, surveilling the ground for enemies likely to shoot up at the troop. He had to move fast to jump aboard, lest risk being left behind.

He barreled into the helicopter, feeling it take off as he positioned himself by the open door. George watched with complete glee as the Kariba base became smaller and smaller underneath him, quickly becoming too far away to see at all. Below him was tall, yellow grass that waved and rippled in the passing zephyrs caused by the craft. George's heart beat in his throat the faster the craft moved, the higher it took him. He sat so close to the door that his foot half poked out of the craft, suspended in air, the farthest it had ever been from earth.

The trip was short. When they landed, George was almost heartbroken as he watched the machines take off, leaving him behind. And yet, he felt better than he had in weeks. That had been the high he'd chased in every run, in every road he'd torn up on his motorbike, in every buggy jump over sand dunes. In the few moments he had to collect himself before more orders were delivered, George watched the pilots fly their helicopters out of sight.

I want to be that guy, he thought. Someday, I *will* be that guy.

But he wasn't.

Not yet.

A mere foot soldier, George's task was simply to patrol.

The idea was to prevent the enemy from coming across the river. The soldiers walked slowly and deliberately under a sun so high and intense it turned their vision to see the world in yellow, the color of the sky blending with the dry grasses that grazed their thighs. Mercilessly hot, George cut off the sleeves of his shirt with a knife and tried to ignore the sting of black flies biting chomps out of his skin. Mosquitos plumed up from vegetation all around him, their slumber interrupted by heavy footsteps. His water supply dwindled within a few hours, no matter how he tried to ration.

On the other side of the river there was always a danger and George felt like a sitting duck in every moment that passed. There was always a bush or two that an enemy soldier might be positioned behind, or tall grass that might conceal a man's crouching body, his weapon aimed at George's head. If the squad was attacked, all men would be reliant on George's machine gun aim. Bitterly, George thought of Robert's advice - *neither first, nor last* – and how that sure as hell hadn't worked out. Now George was first line of defense, first exit from the chopper, last to board. He stuck out.

Then again, Robert hadn't faced foot patrol. Despite his expulsion from high school, Robert was an academic. He'd become an engineer in the gap between school and military training and had earned a more sophisticated position and rank in his first posting. As George stared across the river thinking of how things might have been different for him, his shoulders aching and skin burnt, his ragged clothing sweat-drenched, it occurred to him for the first time why his father might have placed such a premium on education. John Leonard had been required to serve his country as a young man and lived in a nation where his sons would be forced to make the same sacrifice. Perhaps his insistence his children receive private school education wasn't only an effort to improve their social standings or career prospects, but to protect them from the worst of war.

And yet, it was not the thought of a desk job that kept George moving forward. In every terrible step, minute, hour, day, and freezing, frigid night, what gave George spirit enough to keep going was the promise of the next helicopter flight.

Finally, after a week of surveilling the flatlands, the helicopter arrived and George again saw that magnificent view of brutal terrain, and again felt adrenalin race through his veins, and again allowed himself a moment to imagine he was free. And then, he was back at base camp.

The soldiers were given a night of rest in their barracks to prepare for the next day's assignment: a weeklong patrol of the jungle. George freshened himself up and called Hannah from a payphone.

She was thrilled to hear his voice, and George realized then how much he missed her and his life in Umtali. He was too tired to do much talking, but Hannah had much to report, and George was content just to listen.

"Do you ever get chance to listen to the radio?" she asked. "I call in a song for you every night on Serviceman Requests. No one's going to forget you're on duty, not as long as I'm around!" Hannah laughed, her voice sweet and young and unworried. "And I visited your mother to see how she was doing – all is well, don't worry – and I've sent that money order you wanted me to withdraw from your account. I think if you check your balance at the canteen now, you'll probably find you have new credit there."

"I miss you," George said.

He meant it as he said it but didn't quite realize the words weren't only directed at Hannah. George missed all of his life. He missed Hannah, and he missed his mother. He missed Neil and Clive and Jimmy before them, Dad, and Phineas, and Robert and Marion, and he missed Francis, the way George remembered his lost brother from when he was a little boy. George missed Umtali, and he missed the part of himself he had to leave behind before arriving at basic training. He felt a sad, silent, small pang of grief, for somehow it felt as though his prior life was not temporarily put on hold, but simply *over.* It felt as though part of himself was over. Something was happening. Change was in the air.

"I miss you too, George." Hannah said. "Only two more weeks until you're home again."

George nodded. "Yeah."

And then they said goodbye.

~

One night in a real bed did George a world of good. He woke feeling refreshed and in much greater spirits than he had been in the evening before. He was still sore, sunburnt, and bug-bitten, but he was enthused by the promise of the upcoming chopper flight. After all, he had never seen the jungle from the sky.

He didn't have long to wait. The men packed their bags and weapons, and ate breakfast, and then before George knew it, he was flying over the jungle canopy: rich, green, dark.

"They'll hide there," the troop leader said to George, breaking his fascination.

"Sir?"

"There," the man shouted over the noise of the helicopter, pointing to a small clearing in the tree canopy. "If they're down there, they'll be camped out under the trees. But they like to keep an eye on us, watch out for our birds. If they shoot at us, it'll be from those openings. Keep your eyes peeled."

George nodded, looking down into the open clearings with greater interest, clutching his weapon. The machine gun had rested on his lap the entire flight, but George had been so amazed by the bird's eye view he had barely registered its weight.

Of course, the jungle was not glorious on the ground, in its depths. On this tour, the mission was to patrol beneath the thick canopy, ensuring no terrorists found their way to the repeater station that overlooked from the overhead mountaintop. Rhodesia's technology was such that bases in the flatlands had no way of directly sharing information as the mountain between them cut through their signals. To bypass the problem, the first base sent communication up the mountain to the repeater station, and then the station operator repeated the information down to the second base. Protecting the repeater station was a high priority. Failure to do so might result in the cessation of all Rhodesian communications.

The jungle was warm and humid and murky, and while it was the perfect habitat for insects, reptiles, worms, and snakes, it was hell for mankind. The ground was uneven, mountainous terrain, and George was already unbalanced by the sheer weight of the pack he carried. He stumbled often as he trudged on through the mud, slipping on wet earth and tripping over roots that had snaked up out of the ground. The bugs - mosquitos, black flies, all manner of large, noisy, and thick-winged insects – made veils so thick in the air before him it was almost as if he could push the swarms apart with his hands, as though they were gauzy curtains. They flew into his eyes, nostrils, ears. The climate, too, was torturous. When the trees parted and the men walked under sun, they burned. When the triple canopy closed and retained all moisture, dew, and mist, George worried the machine gun might rust. He struggled to hold onto it, the metal slipping in his sweat. At least discomfort distracted from the fact that their enemies might be hidden around any tree trunk.

It was impossible for a man to see more than was essentially just before him. The soldiers walked so quietly it was as if they crept, nearly crawled, and they might have, had other men not been there to lose dignity to. Their world stopped every time an animal or reptile scurried or slithered, audibly snapping a log or branch. The men would flinch and hold their positions, alert faces turning towards the sound, eyes darting. Breath would bate until the lieutenant put down his hand, and then there would be a silent sigh of relief, and then, more of the same walking.

The troop broke their pace twice a day to eat lunch and, later, to set up camp for the night. During their break, they boiled water to hydrate the dried food they packed with them, adding Camoquin tablets to the pot to ward off malaria. They rationed their food and never felt satiated; never for a moment satisfied. During day patrols heat, humidity, and their own sweat soaked their clothes. At night, George shivered and nearly froze as temperatures plummeted. There was news of a soldier who belonged to another troop having been killed during his jungle tour by a lion in the night. The beast had grabbed the man while he slept and pulled him – sleeping bag and all – into the underbrush for a midnight snack. The attack had been so quick the other men didn't realize their colleague had been taken until it was too late. Of course, a fire might have scared off wild animals, but flames would also attract the eyes of the enemy. At any moment, the troop could be shot at or attacked by a band of men greater in number, and then they'd have their throats slit, ear to ear. Any moment could be their last.

Of course, the hell of the jungle made it hard to care.

~

When the week was done, the boys were brought back to the barracks for another night, and again George was greeted by a week's worth of letters Hannah had sent him. He spent the evening showering, eating his fill at the mess hall, and calling his girl back home.

The next day, the men were moved to the city of Kariba, where they were to patrol the gigantic hydro dam – also known as "the wall" – that used the force of the Zambezi River to power three countries: Zambia, Nyasaland, and Rhodesia. The dam was under constant threat; if it were attacked and destroyed, the war would be all but lost as the majority of Rhodesia would be without electricity and unable to communicate.

George had heard this was a favorite assignment, and soon learned why. On this tour, the men stayed in the city, patrolled in vehicles, and slept in beds. They worked predictable two-hour patrol shifts in groups of two, and when it wasn't their turn to patrol, they could shower and rest or even go out to eat in a restaurant. Many of the men indulged in alcohol or *dukkha* joints while in the city as they felt a sense of safety not afforded in the flatlands and the gomos. As Sean was always stationed in the city, George made plans to meet him at a pub on his first evening there. For the hell that was the jungle and the flatlands, Kariba was heaven.

One day while on wall patrol, a commander's vehicle broke down. It was a problem George was familiar with, one his own beach buggy made him contend with routinely. George asked permission to assist and was quickly able to correct the problem. A few days later, George learned he'd received special invitation to obtain a military driver's license. The higher-ups needed someone to fetch gasoline, it was explained, and to perform minor repairs to vehicles, and transport senior officials. George was thrilled.

At the end of wall patrol, the troop's first tour was complete. They were given one week of leave. George paid the same Volkswagen driver to drive him again to Umtali, where he hugged his mother and went to a movie with Hannah and bought a motorcycle the next day.

Except for the money George had spent on dates, or on sending little gifts to Hannah, he had saved up a good sum of money during basic service and his first tour of duty. The Rhodesian army paid him for his service, after all, and ESC topped up that modest payment, so George's income remained the same as when he was laying powerline fulltime. Unlike other servicemen, George didn't spend his cash on commissary cigarettes or beer while on patrol in Kariba, so his bank account had worked up a nice balance. And while he did already have a small motorbike and an unreliable beach buggy, he needed a vehicle he could depend upon to get him back and forth from base. After all, he didn't want his ability to travel to be contingent on the Volkswagen driver's plans forever. In the end, he traded in his gold Honda motorcycle for an Alfa Romero. That bike drove so smooth it felt like George was cutting through the air rather than treading pavement. At first, the salesman didn't want to sell it to George, and said he ought to get parental permission as he was not yet eighteen years old. Then George took out the wad of cash that would pay for the bike in full, and the salesman didn't have much to say after that.

~

The thing that made service tolerable for all foot soldiers was the promise that someday it would end. In Francis's time, mandatory duty for young men had only been four and a half months of training. For Robert, it had been the same amount of training, plus the same amount of time spent on active duty. Still, there was an end date; one most men kept in focus, slowly trudging towards, like the promise of the helicopter flight that kept George steadily walking onward through the jungle. This promise of freedom was important. For some, it was the only reason to keep moving.

George's freedom was set for October 22nd, 1972. At that time, George's duty would be complete and, for the first time in his life, the future would stretch forward without restriction. He would be master of his own time, his own choices, his own fate.

So, George stuck it out. Again, and again, the same scenes repeated: the flatlands and the river were patrolled without incident, although on every tour he fought to maintain vigilance, to not give into routine, knowing that things could go sideways at any moment. And again, and again, those hellish tours were rewarded by a stint at the wall, where at least there were various tasks associated with his military license: gas to pickup, a superior to transport, an engine to repair, and, after each shift, new stories to trade with Sean at the pub.

What changed day to day at home was the news. In the fall of 1972, it grew bleaker and darker. Sanctions increased, and George and Hannah's movie dates were replaced with regular trips over the border to stock up on gasoline. Over the counter radio, there were reports of violence that creeped ever closer to the small cities and seemed truly to be at Umtali's back door. Farms were still being raided, and farmers killed, but now urban buildings were being bombed, and motorists hijacked, women raped and slaughtered on previously peaceful country roads. At grocery stores, housewives' handbags were checked for any explosives that might have been planted before they arrived at the public squares.

Increasingly, the country was becoming divided. Bishop Donal Lamont, the same man of faith that had convinced John and Louise Kelham on the virtues of private school, had a long history of speaking out against the racial injustices committed by the Rhodesian government, and had sided with the rebellion. On his order, the Marist Brothers School opened their doors to insurgents in the dead of night, tending to wounded men, but also housing those on their way to inflict carnage.

One day, only a couple of weeks before his service was complete, all soldiers in Kariba were summoned. George was patrolling the wall in the city, so he and his platoon lined up in the parade square and listened with stone-faced expressions to the news communicated over the loudspeaker. The war had intensified, and conscription standards were to be modified. Men under fifty years old who had completed their service earlier would now be ordered to return to duty for weeks at a time, their "call ups" scheduled on a rotating basis to balance the needs of the economy with the needs of the military. No white male seventeen-year-old would be given permission to leave the country before his service was complete. Going forward, service was to be extended to one year for all new soldiers and all current soldiers could expect to serve for an additional nine months on top of what they had already completed.

George stood with his chin up, throat exposed to the officers in front of him. He gave no indication of the tightening in his throat, the tears he would have let well behind his eyes, the red-hot fury building in his stomach as the faceless voice coolly called out that every man who had already given what was not offered, but demanded, would give more. Any new release date didn't matter. George knew they had him. They had no obligation to stick to the agreed upon time, and even once he was released from his duty, his work power lining would be interrupted again and again by call ups. The military could keep him forever, if they saw fit, and there was nothing he could do about it. To Rhodesia, George wasn't a man. He was simply a possession they owned.

To ease the pain of service extension, the servicemen were given two weeks off over Christmas break. The news was delivered as though it were a privilege, but all knew the opportunity was provided so men could put what little estates they had in order. Things were getting serious.

Later that night, George and Sean met up with the others in George's platoon at the local bar. Every soldier there was devastated. Most drank heavily, and George nursed a lemonade.

"Things must really be getting bad," Sean said.

"So much for the enemy's 'inferior weapons' spiel," another man replied.

As awful as it was, it wasn't active duty that bothered George. He was no stranger to hard work; in fact, he reveled in discovering how he could push his body, how it would consistently become stronger, better able to perform each feat. What bothered George was the lack of control. It was feeling like a puppet, another man's plaything. It was the smug superiority he saw in the faces of his commanding officers – many barely older than he was, often pathetic pipsqueaks who made more money than he did but returned home to their mothers at night while George headed into the jungle with a gun.

Little by little, George stopped giving a damn. Some days his shoes weren't sufficiently shined, or he back-talked a lieutenant, or fought with one of the more smart-mouthed of his platoon. The behavior landed him confined to barracks on days when he was meant to be off. It became usual for him to miss a few days of every tour off because of detention, which increased the anger he felt towards the institution that made such little of his freedom.

He began paying less attention to Hannah when he was home, preferring the solitude of his motorbike, or hanging around with friends, acting wild. He knew his distance hurt her, and he hated himself for it, but nothing changed. He craved action with the same intensity with which he craved sleep in those earlier days. He felt as though his time were running out, and yet Hannah, who had just started working at Salisbury Police Department, was tired by her own long days at work and was happy to settle down for nights spent in. But sitting still made George anxious. Was he living the last days of his life? He loved Hannah, and he knew she loved him, but sometimes when he was with her, he had the same feeling come over him as at training camp: the need to run, to fly. The few days each month he spent in Umtali felt like a withdrawal; it was the only time his body wasn't flooded with adrenaline. He knew he was being reckless, but recklessness was one of the only things within his power.

George wasn't the only man to feel this way. It seemed the entire nation was rebelling against itself in small ways. Communities which had been peaceful were interrupted by public drunkenness and domestic violence. Vehicle accident rates around Umtali began to skyrocket in the usually sleepy town, and people started openly talking about the increase in drunkenness and overall community unhappiness. The town's boys – for they were still boys, only sixteen, seventeen, eighteen years old – returned from their military bases hard-hearted, angry, and detached. Soldiers' deaths were on the uptick, both from violence and accident. Suicide levels increased. Every person George knew had lost someone. They were a nation in grieving. Grieving for their lost sons and, perhaps though not all admitted it yet, for their lost country.

"This war is not winnable," John Leonard told George on his days off.

George was surprised by his father's words. He and John barely spoke, but George knew John, a veteran of World War Two, to be staunchly patriotic.

"It can't be won," John insisted. "Get out while you can."

The idea of leaving shouldn't have been new to George. He knew the military was lying about the fight soldiers faced. Sanctions on Rhodesia increased every day, and whole camps of linesmen were found slaughtered by machine guns. The enemy's weapons weren't inferior, George knew, and their numbers weren't small. Just after the new conscription mandates were announced, Robert had emigrated to South Africa alongside thousands more in Rhodesian diaspora. But leaving the military would mean George had to leave everything: his work, his home, his friends, his mother, Hannah. A future without his home just couldn't be imagined, but neither could a future within it.

George needed a distraction.

And then, George met Precious Miller[3].

The boys had been out on the town in Salisbury: George, Neil, Clive, and Sean. Hannah hadn't been happy about George taking off for the weekend when they had such little time together, but George found himself being uncharacteristically cold to her, a thing he felt both guilty about and powerless over. It seemed nonsensical to be in a heavy relationship when war and his part in it was ramping up. Every day more soldiers were lost, and those on the frontlines, like George, were most vulnerable.

The boys had run into a group of girls while making the rounds at favorite nightclubs for the Umtali crowd. It was Christmas time - the middle of summer in Africa - and days away from George's eighteenth birthday. Even nights were blistering

[3] Name fictionalized.

hot. The guys wore shorts and light shirts, and girls flooded the streets in mini dresses and hot shorts.

The gang had been hanging around on a street corner when the girls walked by. Girls always had a way of looking particularly good when giggling in a group. The guys called to them, and the girls walked over. The kids all already knew each other. They were all from Umtali.

George didn't know Precious very well, but he knew that her family was wealthy enough to send her to the same posh school Marion had attended. The only other thing he knew about her was her mother: Mrs. Miller was the fantasy of every red-blooded male between the ages of thirteen and thirty. Only a couple years earlier, while George was still at the local boy's school, the guys would watch Mrs. Miller as she grocery shopped, the boring task made absolutely fascinating by her good looks. Precious, George realized, was a dead ringer.

But more surprisingly, Precious was interested in George. Before he even realized what was happening, they were kissing. After that, George spent the rest of his extended leave looking for reasons to sneak away, limiting his time with Hannah to visit Precious.

Being with Precious was the action George had been looking for. He went to her like a drunk to a bottle; the infatuation was the adrenaline-charge distraction he'd been lacking for months. But he felt awful at the end of Christmas leave when it was time to say goodbye to Hannah. When Hannah wept and held him and kissed him and told him she loved him, George returned the sentiments with all sincerity. He did love her, that much was true.

George and Sean took the beach buggy back to the military base only because driving double on George's bike got to be uncomfortable during the trip. Before heading off, they stopped for ice cream in Umtali, and, by chance, George ran into Precious. She didn't embrace George – they were in public, after all – but she did pass a note to him, one she had written for him and was waiting to deliver. George read the steamy note as Sean paid for the ice cream, and then George tucked it away in the glove compartment of the buggy to read, again, in private, later.

~

Finally, after many more months of service, George's duty had been fulfilled. He was released from the military until his next call-up, when he would be commanded to serve as part of the Fourth Rhodesian Regiment, or the "4RR," where he would be stationed close to Umtali. Until that time, George could return to fulltime work at the Electrical Supply Commission.

That Monday, George woke early in the morning and packed a bag before driving to camp. He was happy to be back at work, but when he arrived that day, he found the mood darker than he had ever known it. The linesmen and laborers were tense, working without chitchat, seeming paranoid, always checking over their shoulders or staring off into the jungle. Chris Johnson was the first to welcome George, followed immediately by Mrs. Johnson, who hurried to fix him a hot plate of breakfast.

"Where are the girls?" George asked, shaking Foreman Johnson's hand. The foreman shook his head.

"They're with their grandparents," he said. "It's just not safe anymore. I'm thinking about sending the wife home, herself, but then how would I feed the men?"

His gaze shifted over the small camp the ESC crew had made and then into the outskirts of it, the trees in the distance and the tall, yellow grass that could hide man or beast. He raised his hand to gesture at it all.

"Nowhere beyond the cities are safe anymore," the foreman said, darkly. "Even if we're not shot at, the ground could be littered with landmines. The laborers, they're terrified. If we're ambushed, they'd get it as worse as any of us." Foreman Johnson paused for a moment before shaking himself from the low mood with perceivable force. He slapped George on the shoulder. "But what am I saying this to you for? God knows you'd know better than anyone, isn't that right, Serviceman? Come on, make yourself at home."

George thought about the words as he lay in bed. He had heard about attacks on linesmen camps but hadn't given the news much thought. The military spoke only about how those opposing Rhodesia were quickly being squashed, that the enemy's weapons were inferior, that the Rhodesian army's constant patrols and thoroughly executed missions were rapidly curtailing the violent efforts of insurrectionists. But laying there, George realized he'd been a fool to believe their words or to think that he was safe simply because he was out of uniform. His mind flooded with the information he'd received in the background of his life – from the headlines on newspapers held by others in cafes, from his mother's countertop radio, from Hannah's father's new TV. Linesmen were at risk, often shot at by terrorists hiding in the thick of the jungle or blown up by landmines scattered over flatlands.

The only difference between patrol and ESC was the lack of a weapon.

As risk increased for linesmen, fewer men sought employment in the trade. George soon found himself transferred out of Chris Johnson's camp into a crew led by a foreman named Nell. Nell's camp was located near Marandellas, just over two hours north of Umtali.

George's first military call-up came only a couple weeks after working at Foreman Nell's camp. Again, George was back in uniform, this time ordered to patrol Umtali and its surrounding area unless otherwise ordered by headquarters to transport or patrol elsewhere.

One day, George was ordered to the Kariba base, and was surprised to find the area a hive of activity. The number of soldiers that had been brought to camp since he'd been gone had dramatically increased. The average age of the place seemed to have rocketed by two decades with the number of veteran soldiers that had been conscribed back into action, and there was a marked increase in number of Black servicemen on base. They were members of the Rhodesian African Rifles, or "RAR"; an acclaimed Shona contingent that formed in 1940 to serve during World War Two and had grown from one to three battalions during the Bush War. George, surprisingly, found himself on the same platoon as a man named Monders, the husband of Ma Monders who had so reviled George as a student. George was relieved to realize the serviceman had no idea who George was, or how his schoolteacher wife had felt about him.

But a large part of George wondered if the extra bodies were all for naught. Did they give the military a real chance, or was the war truly unwinnable, as his father had warned? Were men in service just to die?

On St. George's Day, April 1973, George's platoon was instructed to provide relief to other soldiers who had completed their patrol responsibilities in Mukumbura, a city on the border of Rhodesia and Mozambique. The task included a three-hour drive through the flatlands, and much of the route snaked through isolated country roads.

The leader of George's platoon was a man name Blythe Stone. Corporal Stone was much bigger and older than the wretchedly young servicemen under his command. The servicemen sat in the back of a canvas-topped military truck with barrels of gasoline tucked beneath their legs, while the radiomen and the RAR drivers travelled up front, in the cab. The platoon's mission was to secure Mukumbura's borders, as well as to ensure the valuable fuel they were transporting was not stolen by enemy troops along the way.

George was assigned to the first truck, which also transported a large stock of ammunition in a trailer attached to the truck's hitch. The day was bright and clear, but the drive was arduous. There seemed to be no signs of danger, but despite travelling at a crawling speed, the driver seemed to hit every pothole and bump on the road, knocking George into Tony, the man who sat shoulder to shoulder beside him, causing their weapons to clang against the steel drums of fuel underneath their seats.

Eventually, the drivers stopped so the men could stretch their legs and take a leak. The stop was made longer by the troop's need to inspect the vehicle for any damage it might have sustained during its violent jolts.

"May I drive?" George asked Corporal Stone. George's temper had made its way to his mouth. "I have a military license, you know."

Stone gave George a look. He was exasperated, too, but George was stepping out of line. "We need you in the back," the corporal said. Each truck required one machinist in the back, and George filled that role in his.

Eventually, the convoy resumed, travelling down a long narrow road with cornfields on either side. A deep ditch ran to the left of the road, with patches of trees shading the right. It was very quiet, George realized. It seemed for a moment that no insect buzzed, no bird chirped, and no breeze rustled in the grass. And then, there was the sudden and unmistakable sound of gunfire.

The truck lost control and George suddenly found himself on the ground and in pain. He'd been thrown from the vehicle. Confused, he searched for the truck he'd been travelling in only moments prior and saw it had come to a violent stop when it collided with the sparse patch of trees that grew in the ditch opposite the cornfield. George could see the truck was in bad shape and was immediately glad he had been thrown out. When the collision had occurred, the hitch connecting the truck to the trailer had snapped and the trailer had crashed into the truck, crunching the space where George had sat like an aluminum can.

Gunfire continued to sound, but George was still confused by the sudden change of events. There were eight rebels standing to the left of the vehicle, just off the road in the tall grass. Each man held a machine gun and wore artillery strapped across his chest like armor that glinted in the sun. The men's weapons were pointed and firing freely at the Rhodesian truck and their occupants. In the crashed truck, a gas barrel exploded into flames.

Suddenly, the rebels stopped shooting and ran. The second truck had come around the bend in the road, George saw, and he realized he ought to get behind it for protection. He scurried from where he had landed on the earth and, clasping onto his machine gun, positioned himself behind the second vehicle. George looked over the tailgate as more gunfire rang out. Before he knew it, the machinist from the second truck was set to work, and in that instant George recalled his training. He lifted his machine gun and shot into the crowd of men, his hand on the trigger, letting bullets rain where they would. The enemy retreated into the bushes where they had been hiding before the ambush. George kept firing, unsure of where he aimed or if his aim would even accomplish anything more than scaring away those who wished to kill him.

Beside George, Corporal Stone steadied himself on the hood of the second jeep, calmly finding in his rifle's sights a grenade held by one of the terrorists. A split second after Stone pulled the trigger, the terrorist's grenade exploded. The bomb meant for George and his fellow soldiers found the man who aimed to throw it, instead. When at last it seemed that the enemy had truly abandoned the battle, George let go of the machine gun's trigger.

Flames fully engulfed the first truck. George ran to it despite hearing a voice behind him tell him to stay down. He knew it wasn't safe, the enemy might have still had their weapons pointed onto the road, or the truck itself might explode - not to mention the trailer full of artillery that was crushed into the truck - but there were soldiers within the vehicle that had not yet exited, and George couldn't leave them to burn.

George found the driver and the radioman of the first vehicle dead, their heads bloody. He knew with deep regret that there was nothing he could do for them; there were obvious bullet holes in their faces, arms, and torso. But in the back of the vehicle, under the canvas tarp, George found Tony was still breathing, although he was on fire. George grabbed Tony and pulled him still screaming from the vehicle, dragging him across the road and into the ditch. The other men helped smother the flames that raged over Tony's body, the gasoline having drenched him when its barrel exploded. George inspected Tony's body and realized with horror that below his left thigh, Tony's leg was only sinew and broken, exposed bone, a bullet having ripped through.

"Medics are coming," the radioman of the second truck shouted. Behind the second vehicle, the other machinist continued to keep watch. "The chopper's already been called in."

Tony screamed again and again. George focused on him, kept talking to him, tried to help him stay calm. Finally, the first helicopter landed, unloading heavily armed soldiers who replaced their bodies with the bodies of Tony and the corpses of the driver and the radioman, and another man who needed less urgent medical attention. They were not far from Mukumbura, George remembered. That was why the helicopter arrived so quickly. They had been less than an hour away from completing their mission before the ambush. George and the other men who were not injured were left to be retrieved by a second chopper. The fresh soldiers would bring the vehicles back to base.

Knowing the sound of the chopper would have scared away any enemy soldier waiting in the trees, George spent the half hour wait surveying the scene, trying to make sense of what had happened. The enemy must have thought the first truck had been travelling alone and weren't expecting the second to join the action. That was why they ran after disabling the first instead of coming closer to confiscate its valuables. He remembered, suddenly, that he had packed a small camera, and felt for the box in one of his pockets. This was the kind of thing people documented, he thought, it was the kind of thing the army would want for information. He took pictures of the destroyed military vehicle, its front wrapped around the tree, and of the burning gasoline, and the broken steel drum that had burst. He walked into the other ditch and found two more bodies, the bodies of enemy soldiers, dead in the tall grass. One man had been so riddled with bullets it was as if his body was joined only as delicately as stamps in a fresh package, and the other had been blown in half from the grenade that he held at waist height. Their weapons laid near their bodies and were by no means inferior. The ZANU and ZAPU were just as armed as Rhodesia was. George documented it all.

When George and the other men arrived in Mukumbura, they were invited to use the visiting barracks, to shower, to change into the clean clothes the base provided, and to eat something. Their mission was done for the day. The next morning George would go on a short leave home.

Released from camp, he visited Salisbury Hospital to see Tony, and to see if he'd live. Escorted down the burn unit's halls by a nurse, George was introduced to Tony's parents, who thanked him for saving their son. Tony's father had the same shock of red hair the young serviceman was known for.

"How is he?" George asked.

"You can see for yourself," his mother smiled, her face all at once seeming puffy from crying, and yet sallow, wasted by grief.

George entered the room to find Tony alert, if doped up by the drugs doctors had pumped into him. He was in good spirits, as good as any could be asked of a man who had lost his leg and suffered such burns. Tony was courageous, stoic, strong.

"I hear you pulled me out," he said, and George realized the trauma must have forced him to black out the events. "Thanks, man."

George nodded. In the end, they all wanted to live.

In the back of George's mind, his father's words echoed.

It's an unwinnable war.

~

After the hospital, George drove to Hannah's sister's apartment in Salisbury. Hannah was staying there on days when she worked for the police department. By the time George arrived, it was late afternoon, and he knew she'd be off work. He walked up the stairs and found he was correct. She, her sister, and her sister's boyfriend were standing in the kitchen talking, and George got the distinct sense they had just been talking about him by the way their eyes turned to saucers as he walked through the door.

"Hi guys," he said, grabbing Hannah's hand. "I'm sorry," he said, pulling her through the open door. "I have to be with my lady."

George and Hannah drove down to one of their favorite spots, a hotel with a nice swimming pool and a gourmet restaurant where the wait staff wore white, starched uniforms. This time, however, George and Hannah just sat in the buggy.

"The call came in yesterday," she told him. "I heard it over the police scanner. I knew it was you. I tried calling you, and couldn't get through, but I just knew it was you." Hannah was crying but George felt numb.

"I was in Mukumbura. I couldn't call." He didn't think to, George realized. "I'm sorry."

"No," Hannah laughed. "No, don't apologize. It's alright. Another call came in later, they reported what happened, and that you were alright. I called your mom and told her you were ok."

George nodded. "Thanks."

"Do you want to talk about it?" Hannah asked, but George shook his head.

"I want to go for a bike ride," he said. "You want to go back to Umtali?"

Hannah nodded, and they went home.

They spent the day together on George's bike, Hannah hanging onto him from behind, until night fell, and it began to get cold. Then they parked at George's parents' house and hopped into the beach buggy because it was warmer. George had to visit Sean's parents. They'd left a note with Louise that there was something they were hoping George would take back to camp for them.

George left Hannah in the vehicle as he quickly picked up the parcel from Mrs. Griffith's home. When George returned to the car, he realized the mood inside had shifted. Hannah was angry.

"What's wrong?" He asked.

She opened the glove compartment and pulled out an envelope and passed it to him. "It certainly seems you and Precious have gotten to know one another well."

George's heart sank. He had forgotten about the note, had forgotten about Precious, to be honest.

"I'm sorry," he said. It felt like he'd said those words all day, and like he might say them for the rest of his life.

Hannah looked him straight in the eyes as tears ran down her face.

"Take me home, George," she said. "It's over."

~

George didn't see Hannah or Precious again after that.

Instead, he went back to his mother's house and tried to eat what his mother put in front of him and ignored Marion's questions about what had happened.

"At least you'll get a medal out of it," Marion chattered. "That'll be something, eh? They might have a parade for you. My brother the war hero."

"Yeah," George muttered. "Maybe. Whatever."

After supper, John Leonard insisted again – louder this time, and without worry of Louise's hearing – that there was no way for Rhodesia to win the war, and that it was no longer Rhodesia's to fight. It was Ian Smith's war, it was Robert Magube's war, but it was not George Kelham's war. It had been a long few days, perhaps the longest of George's life, and he didn't want to hear talk of politics. Still, he knew his father was right. It wasn't George's war, and it wasn't Tony's either, but it had still cost Tony everything.

A few mornings later, George packed up to return to the base.

"Go to Robert in South Africa," his mother whispered in his ear as she hugged him goodbye. "This war can't be won, Son. You need to leave."

George just shrugged. Leaving the army would mean leaving Rhodesia.

"If you and Hannah are properly done now," his mother gently asked, as if reading his thoughts, "Is there as much reason to stay?"

George didn't know how to answer. He just hugged his mother goodbye and went back to work.

Life fell into its similar routine, this time just that much emptier without Hannah included in it. Detached, and perhaps somewhat numb, George patrolled where he was told to patrol for three weeks, and then spent three weeks in the ESC camp for Foreman Nell, working with one eye ever surveilling the horizon, the shoulders of every man on crew raised high, hunched and tense. He would be called up and away from his real life again and again, indefinitely, for as long as men with more clout than he wanted him to fight their war.

One day, George's presence was requested by his lieutenant. George met the man in his office, standing straight with his eyes facing forward.

The officer was angry, but not at George. Towards George the lieutenant was apologetic, indignant on George's behalf. He had recommended George for commendation; he had wanted the great personal risk George had taken to save Tony's life to be honored. But a recent letter from the lieutenant's higher-up proved it was not meant to be.

"I'm sorry, George," the officer said. He read directly from the letter:

Mr. Kelham, as impressive as his actions on St. George's Day might have been, has spent a great deal of time in barracks as discipline for various infractions. He is here as a serviceman, simply on mandatory conscription, and is without official rank. Consistently, this serviceman has proven to be a thorn in the heel of his superiors. At this time, we will not be recognizing Mr. Kelham's actions. The Kelhams of this world do not receive medals from the Rhodesian army.

George just stared straight ahead. He kept his breath steady, pushed his rising anger downward until it was hard and heavy in his stomach, like a cannon ball in his gut. He didn't say a thing. There wasn't anything to say, only one thing to do.

It was time to get the hell out of Rhodesia.

PART TWO

SOUTH AFRICA

June 1973

When the plane touched down, George realized quite suddenly that, for the first time in his life, he was truly a free man. There were no more expectations or duties placed upon his shoulders, no plans to make except his own, no more obligations to fulfill. He could work and make something of himself, or he could lie down in a ditch and starve. The choice was his alone.

In the same passing thought, George realized some small irony: that the only way to become free was to become illegal. George had no right to be in South Africa, but he couldn't stay in Rhodesia, either. He was nationless, a man without a home. Illegal, but free.

After George made the decision to leave the military, things moved fast. He knew he would no longer be welcome to reside within Rhodesia and had to make other plans, so he sold his bike and beach buggy and tore up the note Precious had given him. Next, he had a talk with Phineas, that wonderful, loyal man who'd served his family so well, and asked him if he would like to make more money. Both men knew John and Louise wouldn't require Phineas' services much longer, as the household had rapidly shrunk in number of residents.

"That depends on the opportunity, Mr. George," Phineas smiled.

George smiled back. "Would you ever consider power lining? It's difficult work, but you're young and strong, and you'll earn about thirty times more than my parents can afford to pay you. I'd be happy to contact my employer on your behalf."

Phineas nodded. "Thank you, Mr. George," he said. "I would really appreciate your referral."

The same day, George made the phone call and Phineas made his last arrangements with the Kelham family, having been immediately hired by ESC.

Finally, George kissed his mother goodbye and shook hands with his father who, for the first time in years, seemed pleased with a decision George had made.

George hadn't fled the country. Rather, he'd officially ended his relationship with it; he released his rights to Rhodesia in the same way he'd given up his weapon back to the armory and had his superiors sign off on his release forms. He saw the irony in that exercise, too: if George had had a different last name, or if his father had made more money, George likely would have been collecting letters of referral from the men from who he instead collected signatures for discharge.

While he was sorry to let down his employers at ESC, who had supported him during service and who had hoped he would return to work for them afterwards, there was a mass exodus of young people leaving Rhodesia, and Mr. Cappa understood. Many former Rhodesians had emigrated to England – most often their parents' or grandparents' homeland – and some fled to far flung places like Australia or the United States. Most, it seemed, moved the short distance to South Africa; a nearby place where they could host extended family members who visited on weekend trips, and enjoy the same climate they'd known all their lives. But there was social unrest in South Africa, too. While they weren't fighting a war, apartheid policies were rigid and the country had a long history of racist policy-making – a history that, even in the forties, had been too unsettling for John and Louise Kelham to allow them to make their home within the nation.

For now, George thought, it would have to do.

There wasn't much option, because young George had fallen in love again, and Eva O'Kelly[4] was not ready to leave the continent.

Eva was from the other side of the tracks – the better side. Her father, Dr. O'Kelly, was a local physician at a hospital and the

[4] Name fictionalized.

family seemed massively wealthy. They lived in a huge, beautifully decorated house on a large plot of land. Her mother didn't work but tended to the home and family, which included Eva's older brother and her younger sister, who was the only child of the O'Kelly family born in the Bahamas and not Belfast.

The O'Kellys, as a rule, did not care for George. Irish immigrants, the family leaned left politically and disagreed with the war, of which George had no choice but to have fought. Besides that, the Kelhams were working class and George had not finished high school. It was only George's Catholicism, perhaps, that scored a point in his favor.

But the O'Kelly's dislike for George and the gulf of class in between he and Eva only made the girl more tempting in his eyes. She was seventeen and beautiful, with long blond hair and striking blue eyes. She was fiercely intelligent and, while she was kind, she seemed somehow to have an attitude of detachment and independence that kept George at an emotional arms-reach. It was a hard-to-get game that ended with George completely wrapped around her finger. Still gut-punched from the ugliness that had gone down between George and Hannah, and the part he and Precious had played in it, George didn't want to play around any longer. Discharged from the army and rapidly facing the loss of his homeland and alienation from all friends and family, the more time George spent with Eva, the more he realized how terribly he just wanted a home. That was really all he'd ever wanted – a family to take care of and be loved by. A partner and friend who approved of him, supported him, believed in him. He imagined building this life with Eva and so, despite the few months they had known one another, and without even formally proposing, the young couple began speaking of marriage.

They would not act quickly, or even set a date, but George would begin preparing for their life in South Africa. He would work and save and eventually buy a house, and Eva would study and graduate high school, and then work for a couple years at the post office, where her father had already gotten her a job. The couple couldn't live together until they were married anyway, on account of their religion, and it didn't make sense to marry until there was a safe place to lay their heads at night. George accepted her stipulations on the proposal without exception. He admired her ability to plan, to judge what was best for herself and accept no less.

Spurred on by his need to set up a life for his girl without hesitation, George went to Robert's house after landing in Johannesburg, but didn't stay for long. He immediately landed a power line job just outside the city and, after a few months, had saved enough cash to buy a 900 Kawasaki – a bike with enough reliable power to get him to the town of Empangeni where he could secure better paying work at a shipyard.

As soon as George arrived at the shipyard, he asked a worker who the foreman was, and then he asked for a job.

The foreman asked many questions, including if George had a visa to work in the country. George lied and said yes to everything and the look on the foreman's face told George that he saw through the falsehood, but that he didn't care. Work was busy and half the Rhodies in the country were illegal. All the foreman wanted was a body to do the work, and for that body to answer "yes" when asked if he was legally able to work. No paperwork was necessary.

The foreman used his hand to direct George to a group of temporary trailers. "We'll put you up in a camp room," he said. "Unpack, get something to eat, and rest. Come back Monday morning and we'll set you to work."

George set himself up in a room, and then made his way to Robert's home where he asked to borrow a toolbox. His big brother laughed.

"What do you need a toolbox for?"

"Got myself a job," George said. "At a shipyard."

This only made Robert laugh harder. "Doing what?"

George shrugged. "Guess I'll find out on Monday."

George arrived early to the worksite the next week, full of hope that he was qualified to perform whatever he had been hired to do. George learned that the project was to build a new coal loading facility. South Africa sold coal to Japan, but the country needed more room to store the extra rock coming in. George's job, therefore, was to help build an additional storage facility. To his dismay, George learned he was hired as an electrician, which he vaguely remembered telling the foreman he was. Of course, George hadn't any more to do with electrical work than the power lines he'd helped erect throughout Rhodesia, but he was a fast learner. More than that, he was motivated to learn.

He copied other men on the job, watching their skills in the periphery of his vision while George developed his own. He took as many overtime shifts as they'd give him in order to learn his work faster, and he tried his best to stay under the radar. Within a few weeks, George had learned enough to get by. Within a month, George had received his first pay cheque. With what was left over after allotting cash for food and other necessities, combined with the small bit of savings he still had from selling his vehicles at home, George bought an Alfa Romeo 1750, a little vehicle with a left-hand stick shift.

George loved to drive that car. Having come from a landlocked nation, George was enamored by his new home's coast and how the ocean seemed to stretch to infinity, it's water as bright as a sapphire in the sun. The climate in South Africa was gentler than Rhodesia, all because of that water. It still got hot – *very* hot – but the ocean supplied a flow of air Rhodesia had been denied, and, of course, there was always the promise of a cool swim after a sweaty day.

No longer bound to the military and in such a beautiful country, life should have felt better than it ever had before. George was finally fully employed again and making decent money doing honest work. But the trouble was George was lonely. His friends were still in Rhodesia: Sean was working as an apprentice millwright (and constantly being called-up by the military, as George had been), and Neil and Clive were serving the Rhodesian air force as helicopter technicians. Eva was still there, too, of course. And after the action, adversity, and adventure of foot patrol, and the danger of working in a Rhodesian linesman camp, George now felt dropped into the life of a middle-aged construction worker. His days were a routine of working, driving, eating, calling Eva, watching movies at the local cinema, falling asleep bored in his bed. Often, he found himself daydreaming, not of the future and the life he'd build with Eva, but of the past – of the way the jungle looked from the view of a helicopter and the bliss of being sky bound.

Then one day, in the classified section of a newspaper, George came across an ad for a skydiving school. *Experience the Thrill,* the ad had said. George already had experienced the thrill – well, almost, anyway. He'd never leapt out of the craft without knowing his feet would safely land upon solid ground. But it was something to do, so George registered for skydiving instruction.

Suddenly, his days were filled with far more excitement. George trained in his hours off after work, and he scheduled parachuting lessons for weekends. Before long, he was in the sky again. Not in a chopper, this time, but a fixed wing craft. A plane. Nevertheless, he could look out an open door at the world below him and the humming in his mind – the constant urging to do more, to work more, to become more – simply stopped. In the air, George was free to be fascinated by life, to just be grateful and at peace in the wonder of it all. The sky was more than home, he realized. It was the air in his lungs. It was life sustaining.

More than anything, it was life affirming.

But after nine jumps to his credit, George's parachuting instructor took him aside.

"*Ek is jammer,* George," he said, apologizing in Afrikaans. "But you're just not very good at jumping. You're going to kill yourself, man!" He wasn't cruel in delivering the news and looked every bit as sorry as the words he spoke. Still, George was devastated.

"You're too stiff up there. Too in control. You turn right on the exit. George, I'm afraid you're going to hurt yourself if you keep up with this. Some people just aren't jumpers."

The information wasn't unexpected. George knew he struggled to exit the craft with the same carefree enthusiasm than the true dare devils he'd practiced with so easily embraced. George was not averse to risk, but he liked the danger to be calculated. He preferred flying to free-falling, but without skydiving he'd have neither. No access to the sky meant he was grounded in a mundane existence.

Heartbroken by the loss, George thanked the instructor and turned to leave.

"*Wag!*" the instructor called after him. "George, I don't mean you shouldn't be up there. It's just, if you lean towards wanting control, well… you should be the one flying the craft."

This was definitely something to think about. George remembered that feeling on foot patrol, the first time he exited a helicopter to watch it take off to the sky, leaving him to a week's worth of drudgery. Back then, he had told himself he wanted to be "that guy" – the man able to fly away. He had thought the same when the helicopter took Tony away for medical treatment, and George and the other soldiers had been left behind to wait for the next craft. Further back, in the very recesses of his memory, he thought about the pilot that visited Marist Brothers, and how fascinated George had been with the thought of operating something so complex it seemed to defy all laws of nature.

This, too, was something to think about. George left skydiving school feeling his spirit somewhat buoyed by the idea of registering in flying lessons, but hope was quickly crushed. A quick investigation proved that pilot lessons were exorbitantly expensive in South Africa. It was an opportunity reserved for only the elite; the class systems as rigid there as they had been in Rhodesia. Was George's passion really only to be doled out to him in fleeting moments, as minor reward for offering up his life and body to fulfil some other, richer man's plans? It was too painful to imagine keeping up his current routine for the rest of his life: working at the shipyard, driving around town, or taking in yet another movie, chatting with Eva, eating, sleeping. He knew life would feel better, and mildly more exciting, when Eva joined him, but part of him wondered how long that would last. Would she be happy as a shipyard worker's wife? Would he be happy as a shipyard worker?

Was there any other option?

It quickly turned out there was. Soon after leaving parachuting lessons, a facility foreman cottoned on to the fact that George wasn't a trained electrician, and George found himself out of a job. It was an unfortunate situation, but not long lasting and, as luck would have it, seemed to be a blessing in disguise. George soon found alternate work as a linesman again and was relieved to step back into a profession he knew well.

The pay was good, too, a small increase from his previous salary, and he was able to work seven days a week, spending every moment working towards the life he was determined to build for his well-to-do bride. Soon he had saved up enough money for he and Eva to buy a small home, but then another blow was delivered.

"George," she said, speaking slowly and softly over the phone, "We're just so young. I love you, but I'm not ready yet. I don't want to leave my mom and dad. And I still feel like I just started at the post office. Besides, the war is getting worse, but maybe it's coming to a head. Maybe this will all be over soon, and then you can come back to Rhodesia and we can live here."

George felt his throat tighten as she spoke. Every six weeks or so for the last year, George had travelled back to Rhodesia on a temporary status, just to see Eva. He'd phoned her every night, and worked every day he could, usually only taking Sundays for rest. Every dollar he'd earned and saved, he'd done it happily –

heartened by the fact that all the sacrifice was for her, that she would be as eager to begin their life together as he was. But lately, George had begun to feel as though she'd forgotten he existed. He was heartbroken by her words, but he'd never been the kind of man to pressure a woman or girl to do anything they didn't want to do.

"Okay, Eva," he said. "That's okay, you take your time." He took a deep breath, nodding his head as though trying to convince himself. "I'll wait for you, my love. You're worth it."

On the other end of the phone, he heard the mood in her voice change to one of happiness and relief. She thanked him and told him she loved him, but it didn't make George feel any better. In truth, he wasn't really listening and was grieved to hang up the phone knowing that he was more invested in their future than she was.

Later, though, George realized that with no rush to buy a home, he'd just come into some money. He thought about how he might treat himself. Flight school was still in the back of his mind, but he hadn't saved up anything near being able to afford the expense. Still, he wished for open skies. For adventure, and new experiences. A change of scenery.

Travel would be good, George thought. Somewhere totally new.

He thought of the scenes and images that he'd seen recently at the movies, one of his only sources of entertainment in a nation and city where he knew almost no-one. Shots of desert towns in spaghetti westerns, of snow-capped mountains, of stretches of highway under gentle sun.

America, George thought.

George was going to America.

AMERICA
1975

George didn't intend to blow through all the money he'd saved for his future with Eva. After a few days tending to a bruised ego, George told himself her hesitancy was simply a normal response to the thought of leaving one's beloved homeland and family. He figured she'd come around in due time, but that while he was waiting for her to move to South Africa, he might as well live a little.

The plan was rugged and barely thought out at all. The only thing George was truly concerned about was the cost of a ticket overseas. When he learned a person could fly from Johannesburg to London on South African Airlines, and then from London to New York's JFK Airport on Freddy Laker Airlines for only about seventy South African rand, he was astonished. More astounding was how quickly it could be done; the entire journey could be completed in only three days, nothing at all like the three-week ordeal his parents had undergone emigrating from Southampton to Cape Town. This was the future, George knew. He would cross the Atlantic in only a few hours, at a time when only nineteen percent of Americans had ever travelled by air. Aviation was a miracle.

George had only one connection in America: a girl named Priscilla Lurdens who had visited Umtali as a Rotary exchange student. Priscilla had stayed in the town for one year, and she and George had become friends in the way that most children from very small villages befriend one another. While planning his trip, George wrote to Priscilla and asked if he might stay briefly with her family, and Priscilla, who had become a flight stewardess, wrote with a resounding yes that her parents, Sally and George, would love to meet one of her African friends.

Eva offered George a second connection. Her aunt and uncle, Leo and Phillipa, had immigrated to the United States around the same time her own parents had left for Rhodesia. They lived in Cleveland, Ohio, and would be thrilled to meet their niece's fiancé.

Finally, the time had come for George to board the flight. He flew out of Johannesburg's bright city and into the grey smog of London to make his connecting flight. He then landed in New York City, where the officials granted him a ninety-day visa on account of the many traveler's cheques he had in his pocket. Not staying to take in the sights of the Big Apple, George rushed through the airport terminal before finally collapsing on a grey hound bus seat, feeling great relief and more than some amazement in the long journey he had taken alone.

George spent that first night in America asleep on the Greyhound bus. By mid-morning, the bus had delivered him to Cleveland, where Eva's American uncle picked him up and brought him home to meet Eva's aunt.

Leo and Phillipa were very hospitable and put George up in their guest room for a few days to recover from his long trip and the jet lag that had come with it. They had a college-aged daughter named Sheila, and she, Leo, and Phillipa, were very interested to hear how the O'Kelly's – their Rhodesian cousins – lived. They were very hospitable, and George pondered at his good fortune, happy to be marrying into such a lovely family. Finally, at George's request, Leo drove George and his overstuffed backpack to the Ohio turnpike, pulled over to the side of the road, and shook young George's hand.

George would hitchhike to Los Angeles.

The first stay-over city would be Chicago. It was a long trip, and George had trouble finding a ride at first, but he didn't mind the walk. It was autumn, and the sun shone gently, a stark contrast to where he had come from. Besides, George was accustomed to walking, and had taken more daily steps while on patrol. At least while walking the highway's shoulder in America, George didn't have to worry about stepping on hidden landmines, being ambushed and beaten or killed for the money in his pocket or even just the hell of it, or about being hit by rogue bullets. It seemed that in America, any action George chose to take would either make his life better, if it were a good choice, or worse, if it were a bad one. In Africa, no matter what a person did or didn't do, things would only get bad.

After a few hours of walking and a few failed attempts at hitchhiking, a vehicle stopped for George's extended thumb. The man offered George a ride to Chicago and, once in the Windy City, George found a YMCA and secured a bed for a week's stay. Nearby, he found a pizzeria willing to pay him two dollars an hour under the table to wash dishes. For the next week, George worked at the restaurant by night and took in the sights of Chicago by day.

When he'd had his fill of Chicago, George hit the highway again. This time he was in it for the long haul – he wanted to get all the way to Los Angeles.

His hopes were high that he'd find a ride. Los Angeles was a thirty-hour long car ride away. If George wasn't picked up, the journey would take six hundred and sixty-seven hours by foot – not counting the time a guy would need to stop for sleep or to eat.

But George wasn't perturbed. He'd quit his job prior to leaving South Africa, knowing he'd be rehired when he returned, and no one was waiting for him at home. George was used to sleeping rough since his army days, and had no problem falling asleep under the stars on crown land. He didn't have a lot of money, but he was a wealthy man in one way: George was rich in time.

Fortunately, luck was also on George's side. The highway was abundant with generous drivers willing to drive George a least a couple hundred miles. None asked for cash; hitchhiking was common at that time, and most drivers assumed the young men who walked the highways to be young, harmless travelers, likely broke. Sometimes a driver took George through a drive-through, but most times George simply gave up his ride when he became hungry, requesting to be let off on the side of a street when he saw a fast-food sign. George ate cheap: hamburgers and French fries for thirty cents, whatever quickly quieted hunger pains without breaking the bank. Then he'd head back on the road again, under the wide-open sky and gentle sun.

George knew he was getting close to L.A. when a red Camaro pulled over to offer him a ride.

"You're actually going to let me get into this thing?" George asked the driver, laughing.

The driver shrugged and smiled behind his shades. "As long as you keep your feet off the dash, son."

The Camaro took George all the way into Los Angeles and dropped him near a YMCA on skid row. George was shocked. He had expected the city to be as bright, wide, and clean as Johannesburg – like what he'd saw in the movies, populated by attractive people competing to one day have their name in lights. But while the city was bright, it was also dirty, dingy, and obviously impoverished. People lived in squalor on the streets or in slum towns created by tents and boxes. George had seen poverty in Rhodesia, but no one starved. There was always food to be enjoyed in Rhodesia, fruit grew freely, and wheat was plentiful, and there was always common land on which a person could take shelter, or even nap undisturbed under a tree. In L.A., it seemed there was only poverty. There was nothing green, nothing alive, nothing to eat but what maggoty waste a person might scrounge from dumpster diving. There were no wild animals to worry about, but he quickly realized there were indeed predators: people were mugged, beaten, and murdered right in the city streets. George also saw when these impoverished people rested – on sidewalks or benches – police officers quickly hustled them along. The jungle in L.A. was made of concrete but was every bit as dangerous as the one back home.

In many ways, George had idolized America as a land of freedom, and he had enjoyed a sense of possibility in Cleveland, Chicago, and on the long, vast highway that stretched across the country. But Los Angeles, a place famed for its perfection, had all been a façade. How had he not realized Hollywood would be fake and phony, all just a cardboard set-up? George didn't stay long. He found his way to a Greyhound terminal and learned he could purchase an unlimited travel ticket for a reasonable investment. Disheartened, he boarded the bus.

Then, George arrived in Vegas.

The bus terminal was right on Fremont Street. George couldn't believe his eyes as he slung his pack over his shoulder and stepped into the glittering, noisy, exuberant night life. It was as though he'd walked into a dream: the feathered showgirls, the gold-plated Elvis, the chiming of slot machines that rang out into the street. Music pumped from the open doors of casinos and laughter was everywhere.

George stayed in Vegas for a few days, gorging himself on two dollar all-you-can-eat buffets and sleeping in comfortable motel room beds for little more than a song. All the while, he wished he'd skipped L.A. and spent more time in Sin City instead. After about a week, George reluctantly left that shining city of lights. It was time to begin the two-thousand-mile journey to Grand Rapids, Michigan. He boarded another bus, flashing the driver the invaluable travel pass George was making work double time, and after only a couple of days, he arrived at the Lurden's home.

It was a little awkward accepting such hospitality from Priscilla's parents when she herself wasn't there, but Sally and George Lurdens assured him that they were thrilled to host any friend of their daughter's. A stewardess for Delta Airlines, Priscilla had simply been held up on an international flight and would be home soon. George ended up staying with Priscilla's parents for almost two weeks before she returned home, but they took him in as though he were a long-lost family member. The Lurdens were just as the movies made all Americans out to be: athletic and welcoming and generous. They showed him their beautiful city with great pride, and Priscilla's older brother, a businessman, took George to a Rotary Club meeting. Entrepreneurialism was the American way, after all.

When Priscilla finally arrived, she was thrilled to see George. He updated her on all the news on Umtali – what her old friends were up to now and what state the country was in. With her help, George found a good YMCA to stay at and a way to make a little bit of money. George was once again washing dishes and getting paid under the table, a job he started in late autumn and kept for six weeks, celebrating his twenty-first birthday with his hands covered in greasy suds.

But Grand Rapids was so beautiful that George seriously considered staying. The country was prosperous and exciting. Best of all, it was peaceful. George had made a friend at work – a large Black man named D.C. Suing who also worked as a dishwasher – and the pair passed their shifts joking around or chatting about what they'd do with their riches once they started making some actual money. D.C. was a few years older than George and already married, and often the guys' conversations would drift into talk of their domestic goals as well as professional: home, children, someday maybe even grandkids. D.C. was all praise for his new bride, and George told him about Eva and the life he planned on building with her.

Each day, George marked off one less on his ninety-day traveler's visa and debated with himself on whether he should board a plane. He was, after all, just as illegal in South Africa as he was in America, and perhaps Eva would find it easier to emigrate to the USA, where at least she had an aunt and uncle to claim. But in the end, George decided it wasn't worth it. If his alien status was discovered in SA, it would be a slap on the wrist. If he was found out in the US, it could result in a lengthy stay at a deportation facility, and he could be denied re-entry into the country for life. He didn't want that. George knew in his heart that somehow America would be his future, so he'd better take it seriously.

Besides that, George wasn't making nearly enough to live well on, never mind support Eva on, too. If he had legal status, it would be so much easier to find a decent paying job. He was getting tired of scrimping and saving. Grand Rapids was only a couple of hours away from Detroit – *Rock City* – and yet George found by the end of the week he could never afford bus fair to catch a concert. It seemed like he was only steps away from Pontiac Stadium, and yet he missed The Who, Elvis Presley, and many more great acts. It was frustrating to be so close and yet so far away from greatness, but it gave George goals to work towards.

Someday, George promised himself, he'd go to any concert he wanted to. Someday, he knew, he'd have a life in America. Someday, he swore, he'd have citizenship in a peaceful, bustling country. He'd have a family, a career. Wealth. He'd belong, and he'd be somebody.

A week before his status in America switched from traveler to illegal, George booked a flight to South Africa. With new dreams and a heart that had only grown fonder by separation, George was eager to tell Eva about all he'd experienced, and to convince her what riches they might find together in the United States of America. The country was theirs for the taking.

LONDON
1977

The apartment was a hovel.

It was one room stuffed with two beds, a hotplate on a sideboard, and a sleeping bag to roll out on the floor when all three flat mates were home and there were more bodies than beds. George lay awake on one of the paper-thin mattresses, listening to the cold London rain slash against the side of the seemingly ancient brick tenant building.

For the first time in his life, George was still. This was a great contrast to his usual self, but everything was a contrast to normal these days. All his life, George had known only the sun, but here there was only rain. In childhood, George had known heat, but in Britain it was only cold. In Rhodesia, everything was new: buildings, roads, families. In England, it was as if things had lasted forever.

In Rhodesia, nothing lasted at all.

The first thing George had done after returning from America to South Africa was call Eva on the telephone. It was late, but he expected she'd be home and she knew he was travelling. Surely, she would want to know he had arrived safely.

As he dialed her number, he could almost see her in the little apartment she'd rented in Umtali now that she was out of high school and working for the post office. He imagined her in a cotton nightgown, perhaps reading a book or a magazine in bed, and he imagined her cradling the phone receiver in two hands when she heard his voice on the other end. He imagined - a fantasy now, he realized - the relief she'd be flooded with, simply by the knowledge that they once again shared the same continent.

The imaginings were quickly dashed upon the sound of Eva's voice on the other end of the line. She was hushed, her side of the conversation dull and distracted. Her answers to his questions were summed in only one or two words, as though she didn't want whoever else was in the room to overhear or guess who she was talking too. George got the sense something was wrong. George asked if she was seeing someone else, and she told him she was.

"It's over, George."

It was news George couldn't accept. He insisted she meet him to talk, or at least say a proper goodbye. They were engaged, for god's sake.

"If I come to Umtali, will you meet me at the Wise Owl?" George asked.

Eva reluctantly said she would.

She had sat across the table from him, arms crossed as she nursed a gin and tonic. George barely touched his lemonade. He told her he could forgive the affair. He had been gone so long, after all, and she was young and beautiful – of course mistakes would be made. But watching her across the table, seeming so cold and bored, George felt he barely knew her. Who was this cruel woman?

"I thought we had a plan," he had said. "I've been working six days a week for us, Eva. For our future. I spent next to no money in the States because I was saving what I had made for us. If you just came to South Africa, I really do think you'd be happy. I have enough to buy us a little house, or we could keep renting until I save more if you need something larger. We don't have to have a big wedding. You could just come over. We would figure it out."

Eva had just shrugged.

"Who is he?" George asked.

She sighed. "He's a lieutenant, George. His name is Warren. My family likes him."

George nodded, bitterly. "A lieutenant, eh? I thought you and your dad were against the war?"

Eva shrugged again. "We're just better suited."

George paid for their drinks and left the pub. He spent the night at his parent's house, feeling untethered and heartbroken and alone. Out in the world without a bond to Eva, to Hannah, to his parents, or to his country, George felt almost as though he was free falling, although without the thrilling majesty of the sky. He had no control over this part of his life, and no plan, no idea what to do.

He'd tried to tell his parents about the break-up nonchalantly, but they didn't take it that way. Louise was visibly crestfallen. She'd had high hopes for George with Eva, not only because the O'Kelly family was in high standing, but because they were also Catholic.

"So, what now?" John Leonard had asked. "Can't hide out in a shipyard the rest of your life."

George ignored him.

"He won't," Louise tutted at John.

"You know, George," she said, gently, "A lot of kids in Umtali are going to London. Back to the old country," she shrugged, as if poking fun at the thought of the children of English immigrants returning to their parent's origin. "Poor Francis, God rest his soul, that was his plan, too, wasn't it? In fact," she said, standing, "the Winces' boy has gone over."

"Not Jonathan!" George exclaimed. "If Jonathon's there, more reason for me not to be!"

"No, no, no," Louise said, laughing. "Mr. Wince's youngest, Sebastian. The lovely one. Mr. Wince was telling me all about it at work. He said Seb's having a jolly time tending bar, living with a mate, seeing the sights…" She rummaged through the family junk drawer for her address book and picked up the phone.

"I'll get his number for you, will I?"

Before George could tell his mother to sit down, she was speaking into the receiver.

"Hello, Mr. Wince?" She asked. "I have my George here, and he's thinking about taking a trip to England. Can I get your Seb's number for him? It might be nice if they could connect."

In the background, George mouthed to his mother to stop, but it was already done. She'd written the number her employer had given her onto a scrap piece of paper and pushed it over the kitchen table to George.

"There. You have it now," she said. "Just have a think about it. A new adventure."

Reluctantly, he took it, thinking he'd probably never call.

Later that night, after John Leonard had gone to bed, Louise Susannah sat beside him on the sofa.

"George," she said, quietly, "I'm afraid I have some bad news, myself."

George nodded, his eyes glued to the black and white television screen his parents had been able to save up and purchase after he and Marion finally left school. He assumed the news would be about one of his siblings, or perhaps that Dad had a heart condition, or was considering retirement – something that would take center stage in the otherwise uncomplicated routines that made up his mother's life. But a mood settled between them that seemed thick and serious. Sensing this change in the air, George turned his head to face his mother and was instantly transported to the terrible place of fear and grief that he'd experienced the morning Francis had died.

"What's wrong, Mom?"

His mother gave a slight, quivery smile and turned her head to one side. "Well," she said, her voice so thin it seemed almost as fragile as a spiderweb, "they say it's cancer, Love."

A chill settled over George's body. "Dad?"

Louise shook her head. "Me."

There was nothing else George could think to say, no words that would make sense of what she was saying, or make him feel better, or make her feel better. There was no language to inquire further into the condition. George didn't know anything about cancer, and his mother's health was such a personal thing. It wasn't a thing anyone ever talked about in their family. It could be an intimately personal question, perhaps about a female part of the body instead of an arm or a leg or a bone. Speechless, George just sat and looked at his mother.

"Mom, I don't know what to do."

She smiled and patted his arm. "Well, you're not going to worry about me," she said. "I'm working with a nutritionist who will get me right back on track. We're so lucky to live here," she continued. "All the fruit! The vitamin D! I'll be right in no time."

But you've always lived here, George wanted to say. If you've always lived here amongst the sun and the fruit, then why would you be sick in the first place?

He didn't ask the question, though. As soon as his mind had conceived it, he realized the truth: his mother wasn't going to fight the cancer. She was simply going to eat and enjoy the sun. She, like Rhodesia, wouldn't be here forever.

And with both gone, so too would be any resemblance of

home.

The next day, the same day George left Umtali, another killing had been reported in the newspaper. A linesman camp was ambushed, and its foreman had been beaten and tied to a power pole he'd helped erect. The man's genitals had been cut off by a machete and shoved into his own mouth. He had bled to death, tied to that pole.

George had known the camp. It had been the last he'd worked at. The victim was Forman Nell.

George left the country knowing it was only a matter of time before these atrocities weren't only happening outside of Marandellas, or in the fields and homesteads of rural farms, or on the border of Mozambique. The war would wash into Umtali. The point of the new regime wasn't equality, and it wasn't only to flex power. It was to inflict terror; to make life so frightening no citizen of the rapidly changing nation would dream of speaking out against the upcoming rise of communism or the new governments brutal tactics or the property confiscation that would soon devastate the nation once known as Africa's breadbasket. No part of Rhodesia would be spared violence. Of course, war would arrive in Umtali.

In March of 1976, while George was packing up what few belongings he had in South Africa to move to London, England, it did. Bombs dropped on Umtali's residential streets, grenades were tossed in windows, and George's former neighbors heard gunfire in the distance. In the months that followed, many people would be injured or killed, the suburbs often ambushed at night. Among the first deaths was a white woman who was murdered in her home, and a black man who was killed when the bus he drove was hijacked.

There would be many more to come.

George was thinking these dark thoughts when he heard the fellows in the hallway outside their flat; their clumsy young footsteps, the way their elbows and shoulders bumped against the narrow hallways built for smaller men, growth stunted by lack of sunlight.

"Georgie!" Steven exclaimed as the door opened and a rush of cool air filled the stale room. "Georgie, you ready to rock or what?"

Steven was a tall, lanky boy with a shock of bright red hair. He seemed to live entirely off Guinness beer, never spending a dime on anything of sustenance. Though having been in a low mood since arriving in London, George had found Steven immediately relatable. Steven, too, could be considered a boy from the wrong side of the tracks. Like George, he had not achieved in school, and had been making his own way in the world from a very young age.

Sebastian, the son of his mother's photographer employer, was gentler than hard-knock Steven, but George appreciated the generous nature and optimistic good-heartedness he'd come to see in his new friend. Seb, as he was known, was more academic than George and Steven. During the war, he had served the military as a police photographer, visually documenting gruesome war scenes and the outcome of spontaneous violence. Seb was slim and wore glasses, and he had an unworried, open sort of face that made people like and trust him immediately. He dressed in smart fashions such as wooden clogs – the "in thing," compared to the sneakers George and Steven preferred.

"Come on then, lazy bones," Seb said, changing out of a rain-soaked jumper into something dry. "You're coming to Earl's Court, and we're not taking no for an answer."

George tried to ignore the lads, but Steven simply picked up and overturned George's bed, so he had to get up or fall to the floor. The only time George had gone out in London was the same night he'd found Seb, two weeks prior, and it had been a disaster.

George had stayed with his father's sister, Aunt Ena, upon first landing in the country, but only for a couple nights. His mother had given him the address to Aunt Ena's small terrace home, and he'd simply taken a taxi to her doorstep and rung the bell. The door was answered by a stylish girl around George's age. She was a cousin, he'd learn, and her name was Sarah. After George explained who he was, he was warmly led inside the old, narrow townhouse, its rooms made bright and warm with floral prints, the aroma of hearty English comfort foods, and the wonderful women who called the place home.

The ladies were thrilled to meet "Uncle John's youngest, all the way from Africa!" They stuffed him full of roast beef and roasted root vegetables and sweet trifle for dessert, and then put him up in Aunt Ena's spare room with a hot water bottle to stave off the chill of the wet, rainy air. George slept for what felt like ten days but turned out to only be two.

When he felt somewhat new again – though still heartsick over Eva and over his mother's cancer diagnosis – George thanked his aunt for her generosity and promised, upon her insistence, to return every week for Sunday dinner. Then he asked for underground train instructions and ventured into Earl's Court to find Seb.

George knew Sebastian, but not well. While they had not been particularly close during their school years in Umtali, George knew Seb to be a stand-up guy, as friendly and kind as his father had been while taking Hannah's photograph. George had spoken to Seb once since then, dialing the number Louise had procured for him. Now far from home and eager to make a collection of familiar Rhodesian faces in the strangely anonymous city of London, Seb encouraged George to find him and promised him a flat to share while he got on his feet. When George was ready, he could find Seb at the pub where he worked most evenings.

George arrived in Earl's Court feeling displaced, unsure of the enormous city and unfamiliar with its many customs. He heard a woman curse in public and was shocked beyond belief. He met a street vendor performing magic tricks and had almost keeled over when the Black man spoke to reveal a British accent. There were so many people, and people from all walks of life in the same place. It wasn't like anything George had ever seen before.

When George found Seb's pub, the place was absolutely rollicking. Seb couldn't be pulled away from thirsty patrons to greet George but shouted above the noise of drunken conversations to tell George to sit and have a drink and wait for the busyness of the place to slow down.

"Go to them, George!" Seb had said, pointing to a table of young Black men sitting in a corner. George looked over.

"Who are they?" he asked.

"Rhodies, man!" Seb exclaimed, handing George a lemonade. "They came in for a brew and I recognized the accent. Their eyes nearly popped out of their head when I answered in Shona. Friends for life now, I tell ya!"

George nodded and thanked his friend for the drink. He made his way over to the former Rhodesians and introduced himself. They invited him to the table, but George's interaction soon proved not to be as friendly as Seb's.

Although the conversation started well enough, all men at the table soon realized they and George had fought on different sides of the Bush War. The men told George he was the oppressor; George told the men they were terrorists. All friendly feelings soon evaporated between them. It was too soon for peace talks between young men. Too many hearts had been broken, too much trauma resided in their bones. At the end of the day, every young man at the table was one in exile of their former home, but anger had not yet given way to grief.

The argument turned into a bar fight that a bouncer had to break up. The group George had sat with left the pub. George stayed at the bar until Seb was off shift, at which point Seb's manager told the lads to leave out the back door to avoid running into the other group. No such luck. A second fight broke out on the street and the police were called. George and the man he had fought spent the night in a jail cell.

Disaster.

"What would I go out for?" George moaned. "To sit on some wet barstool and watch you two get shit-faced?"

Seb laughed. "No, George. You'd come out for the fun of it! For food, maybe? For a few laughs?"

"For girls, George!" Steven roared, punching George in the arm. "The only reason why anyone goes out!"

George shrugged. He wasn't interested in meeting anyone, not after the way Eva had treated him. But he *was* hungry, and bored.

He allowed the lads to take him down to a pub called The Bridge House. The place was an absolute hive of people, the air thick with smoke and the smell of ale. Under George's feet, the floors were sticky, as were the sweaty young bodies crammed so close together within the small tavern. Seb got he and Steven a beer, plus a coke for George. The tables were full, so they stood by the empty dance floor. George, not yet ready to so much as look at another girl, watched the musicians set up their instruments on the empty stage while Seb and Steven talked up the groups of young women who approached the bar to order drinks.

One of the loneliest things in the world is to be in a low mood in a crowd of happy people. George felt miserable. He was desperate to return to the tiny flat where at least he could be miserable alone. He threw back his coke and was almost about to tell his friends he was going home when, suddenly, a new sound filled the air and stopped him in his tracks. He, and every other person in the club, turned as one massive wave to face the stage.

The music was heavy and loud and angry. The band was too, or at least they looked it in their black, torn clothes, the metal piercings in their nose and ears and even cheeks shining under the house lights.

Bad booze will be the cause of your death! the lead singer screamed.

The crowd banged their heads up and down as if emphatically nodding yes. Some of the young men on the dance floor jumped around so furiously that they crashed into one another. Young women held their drinks high up in the air and swayed, their long hair brushing over fitted striped blouses, their legs strong and set firm upon the ground under bell bottom jeans and atop platform wedges. Seb looked at George and nodded, his face lit by the colored lighting that spun over the club and, by Seb's expression, George realized that he, himself, was smiling.

"Not bad, eh?" Seb asked. "Every night, man. A new act every night."

George felt the smile stretch across his whole face, almost hurting his cheeks. Finally, a sign of life! George had always loved music but had rarely ever heard it but through a radio. In sleepy Umtali, there were no shows to attend and while he had bitterly wanted to take in the music in Detroit, Rock City was just too expensive to travel too. *This* was what London was about, he realized. England was in its own sort of revolution, one masterminded by young people in ripped clothing without a penny in their pockets. It was the advent of punk, of new rock, of heavy metal. It was a new war, this one hellbent on fighting classism. Now that was something George could get behind.

Resuscitated by the music and the energy of other young people, George found himself nearly unable to even remember why he'd ever been so infatuated with Eva in the first place. He was twenty-two years old, for god's sake, and girls were everywhere. He didn't want a house or a family, not yet. He'd be crazy to want that now. For the first time in his life, George was a kid – just one of the lads! He was both free and legal, welcomed into Britain as a citizen by right of his parents' birthplace. For the first time in his life, there were no rules, no responsibilities, no duties, no secrets. Life was a game, and the lads knew how to play it.

Poor as poor could be, George, Seb, and Steven scavenged for scraps and were thrilled with whatever they found. They made do and they did without, simply thrilled to be in the bold, underground city, to be part of a new movement, of a new scene. They worked all sorts of odd jobs at odd hours. Steven got both Seb and George a part time gig helping him move furniture, and so for a while the three of them spent their days grunting under solid wood dressers and large sofas or sitting elbow-to-elbow in the front row seat of the moving truck. Working and living together, the mates got really close. Every meal was communal and every night they lay one beside the other, beside the other: Steven on one twin sized bed and Seb on another and George in a sleeping bag on a matt on the floor in between.

To save money for club cover charges, the guys even came up with a shower routine. One pound bought the guys eight minutes of hot water in the building's shared bathroom, but Seb and George soon realized they only needed four minutes each. The lads pooled their shower money, and George would wait outside the curtain with a timer while Seb enjoyed his four minutes in the clean water, hopping out just in time for George to hop in for his. This worked well until Steven figured out how to jig the machine so it spit out the coins others in the building had contributed. Then the guys were able to soak in the luxury of eight-minute solo showers whenever they wished.

Within a couple months, George and Seb left Steven's furniture moving operation and got jobs at the Ford factory in Dagenham, in east London, where George was scheduled to work weekday days and Seb was put on nights. This increased the boy's pay cheques, but also their standard of living. With one flat mate away from the apartment at all times, George was able to move up off the floor and into Seb's bed. He slept there during the nights when Seb worked, and would return to the floor on the weekends, when Seb was home. Returning home from nightshift on weekday mornings, Seb would make George a lovely glass of warm milk on the dresser hotplate before assuming his sleep in the bed George had just left and George, upon returning home from the Ford Factory's dayshift in the evening, would do the same for Seb.

Their food costs were also almost halved by the kindness of a good woman. George visited Aunt Ena's home every Sunday evening, where she took great pleasure in pulling out all the stops for the grown kids she was so proud of. George, Sarah, and Seb would crowd Aunt Ena's perfectly kept floral sofa with their energetic young bodies, their mouths watering as the home filled with the aroma of Sunday roast: beef or chicken or ham with potatoes and Yorkshire pudding, spotted dick for dessert. That wonderful woman felt it was her life's mission to fatten up the young adults in her charge as best she could, and weekly George and Seb were sent back to their flat with as much food as they could manage to balance on George's small motorcycle. It was usually enough to feed all three boys for a few days.

But that didn't stop them from eating out. After all, George and Steven knew how to get the best deals. On one discount evening, George and Steven suggested the lads splurge at a fine restaurant to reward their hard work with a little indulgence. Seb, who had grown up in a gentler sort of life, had no idea what was about to occur.

They'd had no trouble getting seated, even though they certainly didn't look like the establishment's usual patrons, the boys with long hair and crumpled t-shirts as opposed to suits and ties. But despite disparaging looks from the maître d', the lads filled their boots; ordering large, extravagant dishes, and indulging in expensive desserts. When they finished, each boy leaned back in his seat, satiated almost to the point of discomfort. After a moment, Steven winked at George.

"Time for a cig, boys?"

George grinned and nodded.

They stood up from the table and moved nonchalantly, feeling the waiter's eyes follow them towards the door. When they were almost at the exit, George and Steven ran.

"What -" Seb cried.

"Run, Seb!" George shouted as the waiter began to chase them, hollering at them to stop.

George went left, Steven went right, and Seb kept straight.

They ran like hell.

The natural meeting place was in the underground train station. George arrived there first, hopping over the turn style to avoid paying his way in. He was a strong runner, having had kept up the exercise regime since his time in the military.

Steven was next to arrive. He and George sat on a bench and laughed about their luck, their bellies still groaning and almost distended with food. After about half an hour, Steven wondered aloud where the hell Seb was. Another hour after that, George and Steven decided to go home without him.

They hoped to find Seb at the flat and were unhappy to find

it empty but for the beds and floormat and hotplate. Steven and George didn't know what else to do but wait and worry and hope he would soon show up. They each lay silently on one of the parallel twin beds. At a quarter to twelve, they heard footsteps in the hall and sat up straight. The door swung open.

There in the doorway stood the angriest Seb they'd ever seen.

"You fuckers!" He cried, pointing at them.

George and Steven erupted in laughter.

"What happened?" George asked.

"These fucking clogs!" He said, kicking the wooden shoes off his feet so they rolled under the bed. "I tried to run, but the staff caught me!"

Seb had spent the evening washing dishes to repay for the stolen meal. George and Steven howled.

Seb refused to ever fine dine with George and Steven again, so instead they leaned into their routine of working and sleeping and then taking to the wild, old, and yet somehow new streets of London. They patronized loud rock clubs and watched punk acts soar to international fame, seeing the musicians on stage just weeks before they turned up on the covers of Rolling Stone Magazine. They caught Status Quo, Bad Company, and more. At another venue, they caught Black Sabbath and George thought he had died and gone to metal head heaven.

Then, George ran into Pixie.

She was dancing with friends, her honey blonde hair filtering light from the stage. It was Seb who noticed her first, recognizing her as a girl from Umtali. In those days, it seemed as though everyone under the age of twenty-five had fled town for the United Kingdom. Pixie, who had attended the same school as Marion, Precious, and Eva, was thrilled to see the boys, and George quickly realized that she was especially glad to see him, despite the fact they hadn't known each other that well in Rhodesia.

Before he knew it, Pixie and George became an item, but the relationship was nothing serious, just fun, light, and breezy. It was the first time George had fallen into "like" with a girl; not life-changing love, or life-ruining lust, or an infatuation with an end goal to marry. Pixie was just *nice*. She was kind and bright and fun and friendly. She liked to dance and laugh and not think too far into the future. That was exactly the kind of friendship George needed.

Life, as grungy and cheap and unpredictable as it was in London, seemed almost perfect. Having saved a bit of money, George moved out from Seb and Steven's place and rented a room all his own. And then, that old familiar feeling happened again.

Work might have had something to do with it. George was a hard worker but disliked unfair and nonsensical rules. In his eyes, if the work got done, the work got done. And until the work was done, there was no reason to stop unless one's energy was spent. But at the Ford Factory, which was heavily unionized, workers were forced to take an hour off each teatime. The bay doors would open, and canteen ladies entered pushing trolleys with tea and coffee and little pastries down the aisle, and everyone would stop and sit and eat, or go outside to smoke. George found it frustrating. Why stop at four o'clock when you could simply work through the hour and leave early? But that wasn't allowed. George had no choice but to stop and sit and sip and eat – unpaid – for no good reason at all.

He was similarly frustrated by the inconsistency of his work hours. In the six months George worked at the factory, workers had gone on strike three times, tipping forklifts upside down as they left the factory, breaking equipment so their empty positions wouldn't be filled by scabs. What was it all for? George often wondered. When he couldn't clock in, he didn't get paid. He would rather work than strike.

One day, George was moved off the assembly line to another part of the factory where he was told to assemble small parts into a whole widget.

"Your quota is fifty items per day," his supervisor told him. "We expect no more, no less, but you must achieve fifty."

George shrugged. The other people in the department moved slowly. If they were making fifty widgets a day, he was pretty sure he could, too.

George put his head down and went to work. His calculations were correct; the other staff took much longer to assemble the items that George completed swiftly and with great care. Every few minutes, they stopped to stretch and gossip and yawn and look at the clock, but George kept his head down, hard at work. At ten am, only a few hours into George's shift, he'd completed his fifty items. He picked up his coat and got ready to punch out with his clock card.

"Excuse me!" His supervisor shouted. "Mr. Kelham, where do you think you are going?"

George wiped the sweat off his forehead as he turned around.

"Home," he said, gesturing at the table. "I'm done for the day. There's your fifty widgets."

"You're here until your shift ends at six, sir." The supervisor said, exasperated. She laughed. "Just imagine how many you'll have completed then!"

It wasn't fair, George wanted to say. Why should he have to stay and do more than triple the work for the same pay when he'd already completed what the others would take one day to achieve?

In the end, George stayed. He didn't want to lose his job. But the interaction had put a sour taste in his mouth. There was a great scene in London, that was more than true, but it wasn't home, and it wasn't really free. And there was little adventure in it: all of London had already been settled and paved for centuries. There was no wilderness and, economically and socially, little upward mobility. George missed the sun. He missed the sky. He missed fresh air, and not being wet. He really missed flying and knew he could never afford to make that dream a reality in England. He thought back to the last time he felt that sense of open freedom he longed for and remembered walking on the gravel shoulder of an open highway, the gentle sun above him and not a care to carry but the pack on his back.

America.

George had never been a man to wait on a plan. Later that night, he sat at a table in a pub and asked Steven, Seb, and Pixie if any of them would be interested in emigrating to the United States with him. Steven was a firm no. He felt at home in Britain and had a good thing going. The man could have been a poster boy for Guinness; though he'd consumed little but the dark ale during his time in the UK, moving heavy furniture daily had turned the once-lanky lad into a Goliath of a man. Pixie wasn't at all interested either, although she was disappointed that George was leaving. George was surprised. He thought she might be up for the laugh. He looked across the table from Pixie to Seb, who sat opposite him in the loud, smokey pub.

"How about you, Sebastian?"

"I'll go," Seb nodded. "Yeah. America sounds good."

The next morning, Seb and George jumped the underground turnstiles and took a train to the American Consulate in

Grosvenor Square. The plan was that they would simply see an immigration officer, fill out whatever paperwork was needed, and then buy a plane ticket to board a flight a few weeks later.

The U.S. Embassy was in a square surrounded by the embassies of two other nations; Japan and Canada faced the same courtyard as America's. But where the other nations' doors were clear of would-be immigrants, America's were blocked by a long cue that wound out the embassy doors and across the court. George and Seb looked at the line, and then at each other.

"Try again tomorrow?"

"Try again tomorrow," George agreed.

The next day, the lads woke bright and early, hours before the consulate opened. Even then there was a small line when they arrived, but they cued up and waited a couple of hours for their turn.

At last, the boys were face to face with the immigration officers.

"We'd like entry to America, please," George said, handing his passport to the officials.

The Americans looked at George and Seb. Their eyes lingered over the boy's long, scruffy hair, their Chelsea boots, and the rebellion of youth written all over their faces. They slid the boys' passports back across the table.

"Not interested."

George and Seb were flabbergasted, too stunned to do anything but move out of the way when ushered out of line.

"Can they do that?" Seb said. "Can they just say no?"

George looked down sadly at his rejected passport. "I guess so."

The boys stood in the square, dejected and disappointed. *What now?* George thought, and then realized that, standing in that square, the whole world was open to him. As he looked around, he was mere steps from the gateway to Japan, or he could simply travel a small distance and apply to move to Sweden, or France, or Australia. All these doors were potential futures, each outcome presented at the exact same time – all he had to do was choose one.

"I'm going to Canada," George told Seb. The building was close by and there was no line, and George had heard that Canada was simply America's northern cousin, still a little wilder, a whole lot colder, but wide open and young and free.

"Coming?"

Seb shook his head. "Not on your life. Too cold, mate. And too… rural?"

"Too rural?" George laughed. "More rural than Umtali?"

Seb shrugged. "I like cities. I like action. I like heat, man! Two dollars an hour pumping gas in frozen Canada, that's good money, but for what? What kind of life is that? Look, I feel the same as you. London's a laugh, but it will never be home. I'm going to South Africa."

George was disappointed, but no longer shocked. Seb in South Africa made sense. Still, George was sorry the next part of his journey would be spent alone.

George cued up in the short line to ask for entry into Canada and Seb kept him company. The Canadian officials told the boys that as citizens of the Queen's commonwealth, they were welcome to visit any commonwealth country they liked.

"However," the official said, looking their punk aesthetic up and down in the same manner the Americans had, "we do not encourage you to do so."

The boys nodded, and left the line, and then laughed and laughed at the fact that no nation wanted them. What else was new?

In the next couple of weeks, George sold all he had to sell, which wasn't a lot. He emptied his British bank account and kept his money on him in cash. He gave his notice at work and advised his landlord that he'd soon be leaving. He spent a few fun-filled last days with Pixie, and they both felt sad when their time together ended, but neither shed any tears. As she walked away from him down London's streets, she turned back to look once more, and blew him a kiss. In the end, they'd simply been good friends to one another, and that was enough.

George said goodbye to Aunt Ena and Sarah and gorged himself once more on his aunt's lovely cooking. Then he bought a plane ticket to Rhodesia because there was one last thing to do.

George's father had called to tell him to come home and see his mother. It was the first time John Leonard and George had ever spoken to each other on the phone.

George had assumed his mother was grieving the loss of her children along with her adopted nation. With so much upheaval in the country, Robert, just as George was about to do, had moved across the Atlantic to Canada and Marion had left to South Africa.

All three of the Kelham children had encouraged their parents to move as well, but John and Louise felt a trip was beyond them at their age. They had next to no savings, after all. All they had ever worked for had been invested in their children's education, and to leave Rhodesia meant losing John's pension, as meagre as it was.

But when George returned to the house on Rudland Avenue, he saw John had urged him to come home for a very different reason.

The home was empty. Louise Susannah lay in hospital, dying.

George visited his mother in the palliative care ward. She was skeletal thin; weak and bed bound. Her face pinched in pain when she moved, and her hand gravitated to the pain point under her arm as she tried to sit up when George entered the bedroom.

"Hello, George dear," she said, warmly, despite the obvious pain. For a moment, it was as if George was only six years old, opening the door to find his mother waiting for him in the kitchen after he'd spent a day playing with Jimmy Hynes. He sat by her bedside, and she touched his hand.

Louise Susannah told him about the new nutritional plan she was on, and the natural medicines she was taking. She described the healing properties of plants and reminded George about her good strong teeth, and his good strong teeth, and how they had them all because of fresh, wholesome food. Then she asked George about his life, and they talked about Canada, and then Louise asked George to call for a nurse.

"There's something I want to show you," she said to George.

"Yes, Mrs. Kelham?" a nurse asked.

"Could you bring…?" Louise seemed too tired to finish the sentence. She gave a little wave of her hand, instead, knowing the nurse would know what she wanted.

The nurse nodded. "Of course, ma'am."

The nurse retrieved something from a dresser on the opposite side of the room, gently laying the knitted material on Louise's lap.

"Thank you," George said. He looked up to catch the woman's eye, not knowing how to adequately put into words the thanks she was owed.

"This is for you in Canada," Louise said, slowly unfolding the material. "They say it's very, very cold there, George." She had knit him a sweater – a heavy, thick, close-knit black and white jumper.

George smiled. "Well, this is perfect, Mom. Look how heavy

this is! And me without so much as a coat. This will do the trick. Thank you."

Louise nodded and smiled but her eyes were starting to close, and George realized she needed to sleep. It was time for him to head to Salisbury to catch his plane, anyway.

"I'll see you soon, Mom," he said, kissing her cheek.

"I love you, son. You have a safe flight."

By the time George had stood up from the chair at her bedside and taken the few steps to the bedroom door, Louise had fallen into unsettled sleep, her face a mask of discomfort. She made small sounds, moans and whimpers of discontent. George looked at his mother, and all at once, he realized that this would be the last time he ever saw her.

And then, George left Africa for the last time.

Or so he thought.

ALBERTA

December 20th, 1977

George wore the black jersey sweater on the flight across the Atlantic. He carried everything he owned in the world in a bag smaller than the oversized whisky bottle he regretted purchasing for Rob as a Christmas gift. George knew Rob would get a laugh out of the enormous whisky bottle as soon as he saw it gleaming under the lights of a gift shop display, but the bottle was so big it required its own bus seat. By the time George boarded a Greyhound bus outside of LaGuardia Airport, tired and irritated and bogged down by his small carry-on bag, the huge whisky bottle, his passport papers, and ticket stubs, he felt as though he'd dragged the damn thing halfway across the world. In fact, he had.

He'd travelled to New York on a limited visa, and he was given just enough permission to land on American soil and leave for Canada. This was the second time in George's life he'd landed in New York only to immediately board a bus. The city was the epicenter of the world, he thought with some humor, and yet his only relationship with it was to be spit out: first, to Cleveland and, this time, to the great white north.

The bus would enter Canada at Emmerson, Manitoba, and would take George all the way to Edmonton, Alberta. It would be a three-day trip.

On the second day of the trip, George was thrilled to see Canadian signage in both English and French – the country's duel official languages – appear at the roadside. Outside his window, the country was stark and bright and beautiful. The sky was huge, so big it seemed to reach unobstructed into eternity, and it was as blue as George had ever seen, as blue as the ocean off South Africa. The land was perfectly unblemished, white with snow, and it seemed to make itself whiter and brighter still, as though to promote the glory of the sky. As the bus neared the Canadian immigration port, George gathered his things and pulled the whisky bottle close to him, balancing the poison on his hip as he stood to exit the bus. He couldn't wait to greet his new home.

But no one else on the bus moved. There was no usual jostling of bags or vying to get into the aisle before other travelers. Some on the bus looked at George with apprehension. Was he right to stand and ready himself to leave, or were they right to sit and wait? George wondered if he had missed something. The bus doors didn't open. He sat back down.

After a few minutes, a man came aboard the bus.

"Welcome to Canada, folks," he announced. "Please have your passports and visas ready."

George waited patiently at the back of the bus for the man to make his way to him and handed over his documents when it was his turn.

The man, obviously seasoned in matters of immigration and international travel, squinted at George. George was very suddenly self-conscious of his youth and his lack of a travelling partner, as well as his citizenship in a country that teetered on the brink of nonexistence.

"This is a traveler's visa, son," the man said to George. "It's only good for a few months. Will you be looking for work while you're here?"

George shook his head no and did his best to appear believable.

"I'm just visiting my brother," George said.

"Where's he at?" The immigration officer asked.

"Edmonton."

The immigration officer laughed. "Edmonton! In December? Not exactly tourist season, son. But it does seem a lot of young men are interested in Edmonton lately."

The man stared unblinking into George's eyes. Of course, George wasn't the first young man to go west. Alberta offered success to any young immigrant smart enough to work in the booming economy.

George smiled. He'd let the man call his bluff outright, but he wouldn't give himself away.

"Christmas," George said, shrugging. He hoisted the oversized whisky bottle a little higher to draw the officer's attention to it.

The man glanced at the bottle and gave a whistle. "You've dragged that thing all the way from New York?"

"All the way from London, sir," George said. "A present for my brother."

The officer threw back his head and laughed. He stamped George's documents.

"Merry Christmas, son."

George didn't have an opportunity to get off the bus for another day after that. He was glad of the fact that he'd packed snacks and water in his luggage, and that the greyhound had a bathroom. George was pleased enough to look out the window and take in the landscape of his new country. He was happy to listen to the many accents and languages of his fellow passengers as they spoke to one another in hushed tones. Most of all, George was happy to lean his head against the cool bus window and close his eyes to rest, the bright sun warming his face. Jetlag had set in and the time difference had his body feeling wretched. At five PM on the bus, it was midnight in London, and his body was still running on time kept by Big Ben. It was a little bit funny that the bus almost seemed to serve as an incubation zone for would-be Canadians. Travelling in this small space with so many others was a crash course on unassuming curtesy – the politeness this country's citizens were known for – and gave all bodies space to acclimatize to the drastic time changes they faced.

But it didn't prepare them for the weather.

The next day was neither bright nor clear. The wind howled over the flat landscape and snow blew sideways, sending skiffs across crusted fields like sand on the shores of a hurricane. As the bus pulled into a city George didn't know, he was alarmed to see pedestrians bent almost in half as they pushed forward against the wind, their mittened hands clutching the openings of thick parkas, pulling their comically large jackets closer around their necks. It wasn't yet four PM, but already the sky was a dark, inky blue.

Over the intercom, the bus driver announced they were in Saskatoon and that all passengers would get chance to stretch their legs and buy a sandwich at the bus station as the vehicle would be parked for two hours. George lined up in the aisle, packing all his belongings to save them from theft. It was the twenty-second of December, and George wanted to call Rob and let him know he'd be in Edmonton by the twenty-third.

When the bus doors opened, George was shocked by the wickedness of the cold air that rushed in. At the outdoor payphone, he couldn't help his body from shaking against the extreme temperature, his words turning to hot steam in the frozen air. He was too uncomfortable to laugh at his naivety. How had he ever thought a simple knitted sweater would be enough to protect him from this kind of cold?

George's Chelsea boots had finally touched Canadian soil.

"It's fucking cold!" He shouted into the receiver, over the howl of the wind. On the other side of the line, Rob laughed and laughed. George suddenly worried the big bottle of whisky might spontaneously freeze in his arms. Did Rob know if alcohol could freeze? A child of Africa, the thought had never crossed George's mind.

"Where are you now?" Rob asked.

George racked his tired head. "Um, a sign says Says Cat Chewing."

"What?"

"Says Cat Chewing. I'm in Says Cat Chewing!"

Rob howled harder. "Says Kat Chewing? You're in Saskatchewan. Sass-cat-chew-an."

George just shook in his tight black jeans. His bare hand hurt holding the frozen receiver, and he stretched the arm of his sweater up to his palm to use it as a glove.

"Well what kind of name is that?" He laughed. 'God, it's cold!"

"Oh, little brother," Rob said. "You haven't seen anything yet. I'll see you tomorrow."

The next day, Rob picked George up in a baby blue pinto. The car sure looked good with its long engine and its sleek back, but George had heard rumors the model was little more than a death machine. The gas tank, stored in the back of that smooth, streamlined model, tended to burst into flames even at low speeds, often for something as simple as being rear-ended by an impatient driver rushing from behind.

Rob drove George to his home in Clareview, Edmonton. Rob was married to a woman he'd met in South Africa, whom had been married prior to a Canadian. Although she was divorced when she met Robert, her previous marriage continued to afford her – and any future spouses – Canadian citizenship. Because of these lax immigration rules, emigrating to Canada had been an easy task for Rob and Anne. Together with George, the three Kelhams celebrated Christmas with Rob and Anne partaking in the whiskey George had brought while he enjoyed the 'Canadian fare' they'd put out especially for him, knowing he'd never tasted Spaghetti Ragu.

The day after Boxing Day, Robert took George to a retail store and outfitted him with proper outdoor wear. George, combing every classified he could get his hands on, quickly secured an interview for a linesman job only a few days after arriving. Northern Alberta was booming then, and industry would take whoever showed up to work.

Rob and George drove the hazardous little pinto up north to Westlock, and George met with a hiring agent for Fell Pattern, a subcontractor to Alberta Power.

"We're hiring all warm bodies," the agent said, barely asking George his first name, never mind what kind of experience he had or whether he was even legally allowed to work in Canada. "We can put you up in camp or in a hotel – it depends whether we send you to Swan Hills or Fox Creek. Can you start on the second of January?"

George nodded. He would start whenever they would have him. Back at Rob and Anne's place, George quietly celebrated his twenty-fourth birthday and welcomed in the year 1978, and then took a bus north to Swan Hills.

It turned out that laying powerline was much different in Alberta than it was in Africa. In Africa, George's work had been assisted by a crew of ten or twelve and each man completed their task with little more technology than a shovel. In Alberta, however, a linesman was expected to complete each job by himself using equipment and machines that provided some automation of the task. Despite his years of building power lines, it took George some time to figure out how the machines worked. Each night for a couple weeks, he again experienced that bone-tired exhaustion when collapsing into a motel bed, every muscle once again re-inventing itself as his body got used to the new tasks.

Of course, the climate presented its own challenges. One blisteringly cold January day, when George had climbed up a newly mounted pole to configure the lines, he used his mouth to hold a small metal part instead of putting the thing in his tool belt. Once he finished jigging the connector, he pulled the metal piece from between his lips and screwed it into the pole. It wasn't until he climbed down from the pole that he realized blood covered the front of his jacket. Frozen breath obscuring his reflection for a moment, George peered into the rearview mirror of a truck to see where the blood was coming from, and realized he'd half torn off part of his lip, where the metal must have frozen to his face. He hadn't felt it. The cold had made him numb.

Soon George had worked a full two weeks and a cheque was issued to George before his first days off.

"So, is everything okay?" George nervously asked when his employer handed him the envelope.

The man nodded. "All good here. You good?"

George nodded and excused himself, ripping into the envelope as he walked away, his eyes scanning for tax information.

Upon George's hire, he had been required to provide a social insurance number to be added to the payroll, and there had been a moment of panic. George didn't have a social insurance number. He wasn't allowed to work in Canada and was in the country on only a traveler's visa. In a flash of ill-thought-out spontaneity, George had scribbled the digits to Pixie's phone number, a sequence he remembered from his time in London and figured could be good luck. He'd spent the next two weeks on edge, wondering just how quickly he'd be found out and deported. But it had worked. There, on his paystub, was Pixie's phone number and evidence that taxes had been taken off his pay and deposited into someone's Canada Pension Plan and Employment Insurance. George would never again see that money, but he hoped whoever would, could use it.

George hitched a ride with a crew member who was heading into Edmonton for days off and chipped in with gas. There, George passed the long Albertan nights in Rob's basement, trying to keep warm and watching movies on a tiny television set. When called back to work, George was moved all around Alberta, and travelled with the sense of having a target on his back. Meeting new people was difficult and made him nervous. His very voice – his accent, rather – immediately proved he didn't belong. Again, he recalled the advice Rob had given him before he'd joined the military:

Neither first, nor last.

George kept his head down.

Pay came in like clockwork and, every two weeks, he happily handed over a portion to Rob and Anne as rent. George saved whatever was leftover. He couldn't open a bank account himself, so Rob cashed the cheques for him, and George squirrelled the money away in an old coffee can. The plan was to finally make his dream of becoming "that guy" come true. Flight school in Canada was cheaper than anywhere in the world, and the schools didn't check a person's social insurance number – all they cared about was whether you could pay tuition. George told himself he just needed enough to get to school. Pilots were hard to come by, especially in a low-populated country like Canada, and he figured attaining citizenship or even landed immigrant status would be much easier once he was in possession of a sought-after professional license.

There was always a little hum of worry about being caught out in the back of George's mind. He was working in Fox Creek when his supervisor interrupted his work to ask him to step into the office.

"A call has come in for you," the man said.

George was bewildered and panicked. Who would call him? He stomped into the administrative trailer, feeling too large for the small space in his heavy boots and large workwear. He removed his helmet and ignored the way his previously frost-bitten lip stung as his skin acclimatized to the warmth of the room.

"Hullo?" George said, fearing hearing the voice of an immigration officer.

"Hey, George."

It was Rob, but George wasn't relieved.

"I'm sorry to call at work…it's just… I just heard…"

George felt his chest tighten and immediately knew it was bad news. Good god, he thought, I hope I haven't gotten Rob and Anne in trouble.

"When am I supposed to leave?" he asked, but his words were interrupted.

"Mom passed, George," Rob said. "She died last night. She's gone."

George listened to the details, without really hearing them. There was nothing he could say or do. Dad wasn't planning on having a funeral for her. When George hung up the phone, he almost felt as though he should dial his mother's number, thinking surely she'd want to talk about this thing she had just experienced, but then George realized how silly that thought was. Louise Susannah was gone. Dead. He didn't have a mother anymore.

He swallowed the tight lump that had formed in his throat and passed his hands over his face, rubbing his eyes, cheeks, and mouth as though he'd just woken up. He very suddenly remembered the soft way she'd told him about Francis.

He's gone to heaven, Love.

Sadness set over him like a soaked blanket, but George knew it was only final confirmation that he truly was alone and adrift in the world.

But then – no. Although part of him felt lost and alone, another part remembered the way Louise had faded from sight during the train ride to training camp, and George suddenly felt the same sense he had had that very day, as a younger man: his mother, forever holding vigil. George couldn't see Louise Susannah, but maybe, somehow, she was still there – still *here* – her love and prayers still present, always protecting him.

And with that deep knowing, George went back to work.

A few months later, spring arrived. The wind was warm and green grass peaked out from beneath patches of melting snow. But just as George began to think he might like Canada, the camp suddenly shut down. All workers were unceremoniously let go.

"Spring break up?" George asked the other workers as they picked up their last cheques. "What does that mean?"

"It just means we either get some time off, or we find somewhere else to work," a crew member explained. "They'll hire you back on again in six weeks."

George shook his head. "I can't go six weeks without a pay cheque!"

His heart started to beat rapidly in his chest. Changing jobs meant fooling a new business with Pixie's phone number. Each transition meant increased risk, and he still hadn't saved enough for flight school.

But the crew member was generous. He took out a scrap of paper from his back pocket and asked if George had a pen. George handed him one and the man used his helmet as a hard surface as he scribbled.

"Call this number," the man said. "Ace Construction. Calgary. They'll hire you on."

George thanked his former coworker and stopped at the

motel payphone.

~

A representative for Ace Construction told George if he could get himself to Wainwright, they'd hire him the next day. George headed south on the Greyhound immediately and was dropped off at a gas station in the small town. A man named Vik Zipchain was waiting there for him.

Zipchain had the same friendly demeanor and rough exterior George had come to recognize in Albertans. It seemed every third person George met were men on the cusp of becoming millionaires, but perhaps working so hard they didn't even know it themselves. Money was splashed around; there was far too much work for most employers to keep up, and far too few men to meet industry's rapid expansion. Overtime was dangled in front of workers like carrots before hungry horses.

George stood off to the side and his new supervisor got him set up in a local hotel. He watched the girl behind the counter. She smiled warmly at Zipchain as they spoke, and George was immediately taken with how calm and quietly confident she was during the interaction. She was young and petite with bright eyes and curly hair, but her serene and friendly presence made her somehow seem older – cleverer – than her obvious youth. She turned her attention from Zipchain to George to hand him his room key, and George was so lost in his admiration of her that he almost forgot to say thank you. Zipchain gave George the details on the next day's start time, and then they shook hands goodbye. Zipchain exited the small lobby and George left to find his room, turning back for a last look at the girl, but she had already taken leave to the back office.

After a couple weeks of working for Zipchain, George was again relieved to find Pixie Parker's phone number had again done its job. George was making seven dollars an hour working for Ace Construction, a sum that almost made his mouth drop when Vik proposed it. George didn't try haggle the wage higher during negotiations, but afterward he realized he should have, and that likely it had been expected of him. It didn't matter, George was just happy to be making some decent money.

With his first pay cheque, George bought a dirt bike. He was thrilled to find Canada wasn't all cold and snow, as he had been warned. He learned that May and June got so hot that bush fires were a common problem. George took care to maintain his new vehicle so that no sparks ignited grass in the rural ditches and roads around Wainwright, the trails making perfect dirt biking tracks. Soon, George formed casual friendships with other young men on his crew who also owned bikes. He began spending days off with Canadian crew members, ripping up and down dirt roads. In the evening, the other men drank Molson beer at the local watering hole while George nursed a coke, and everybody, including George, hollered at the television set whenever the Edmonton Oilers took the ice.

In his earlier life, George hadn't been much of a sports fan. Any love for physical competition had been spoiled by the mandatory athletics he'd been required to devote all free hours to at Marist Brothers. But there was something different about hockey. Hockey was fast! George admired the way players soared across the ice on thin metal blades and he loved the game's contrast of silence and sound: athletic voices muffled by helmets, but their actions emphasized by the slam of large bodies into boards, the crack of a puck against sticks, a skate's high, thin song as it cut into freshly cleaned ice. He got a kick out of the way Canadians roared while watching the game, the citizens of the cold nation known for otherwise being polite and mild. In Canada, hockey was a great equalizer. It was something to be enjoyed in comradery with the stranger sitting beside you: it required no common language, no shared heritage, no family history. It seemed as if social class didn't truly matter in the new country. In sports bars, George observed laborers sharing tables and conversation with small town doctors, or the town mayor might buy a round for the house and a self-employed plumber could slap him on the back and return the favor. The country's favorite game was broadcast nationally and freely over CBC's Hockey Night in Canada, then only a two-year-old improvement on the show's radio origins. George's initiation into hockey culture was as meaningful as any religious rite he had been expected to undertake. This was the first way George declared himself a spiritual citizen of a country that didn't yet know he had arrived.

But aside from burning gasoline on dirt bike trails or indulging in deep fried sports bar fare, George lived frugally. He ate soup out of cans boiled on the hot plate in his room. Seventy-five cents bought him a full meal at fast food dives. He purchased clothes only when his tore beyond repair during tasks on the construction site. All the while, he put whatever he could into savings. All the while, he told himself that one day he'd become a helicopter pilot.

The whole time, only George, Rob, and Anne knew his secret. George was an illegal alien and deportation was only a matter of time.

In the fall of 1978, George was transferred from Zipchain's crew to work under a foreman named Grant Penalton, a good man who lived on a nearby farm where his wife often hosted dinner for his crew. Penalton reminded George of the friendly, good-natured Chris Johnson, his first supervisor in Rhodesia, and George made sure to buy a couple boxes of smarties to give to Penalton's young daughters whenever he visited the farm.

Under Penalton's leadership, George was given a raise all the way up to nine dollars an hour and felt the money burning a hole in his pocket. He'd already saved up over half of what helicopter school would cost, and figured it was time to use a little to increase his standard of living. He was tired of buses and hitching rides to new job sites. George was ready to purchase a car.

After some consideration, George chose a 1973 Lincoln Continental – secondhand, of course – but in good condition. Driving the huge, boat-like car to the hotel, George felt like a million bucks. His Canadian experiment was working. Life was progressing just as it should.

Well, except for one area.

George was lonely. It was a truth he couldn't deny any longer. He'd made casual friendships at work and was glad of having a small crew of guys to dirt bike and watch hockey with, but those men drank like fishes and George wasn't about that. He had met a girl, too, but she was silly, superficial, and George got the sense on more than a couple occasions that monogamy wasn't her strong suit. Whenever George talked about music, or reminisced about Africa, or fantasized about the future, it was like each conversation took three times as long just to catch her up to speed.

More and more, George realized the best part of his week totaled a maximum of two minutes spent chatting with the girl at the hotel check-out desk. He desperately wanted to spend more time with her but struggled to find valid reasons to interrupt her duties. After all, Penalton was responsible for signing off on all the hotel bills. The only chance George got to talk to the girl at the check-out desk was to request more towels in his room. George started ordering towels so often that she must have thought George showered five times a day.

On the day he bought the Lincoln, though, George felt good. Really good. The drive had given him confidence like nothing else, and when he entered the lobby and saw the girl at the front counter, he decided that it was his day.

"Hi," George said.

"Great car," the girl replied, her warm smile drawing him in. "I saw it as you pulled up. Congratulations."

"Thanks," George said. He glanced at her name tag, hoping she wouldn't notice. "… Deborah." He already memorized her name but speaking with her had disarmed him and he didn't want to screw it up. "I'd love to take you for a drive," he said. "We could go to the movies."

The girl smiled and cocked her head. "That could be fun," she said. "But only if you call me Deb."

~

It was about twelve seconds into their first real conversation when George fell in love. Deb was sweet and grounded and independent and fiercely intelligent in a commonsense way that made her seem unshakable. She was sensible in a way that reminded George of his mother, and yet she seemed fearless in a way that George aimed to be himself.

When he told her about his ambitions to become a helicopter pilot, Deb asked him when it would happen. She wanted to know what date his schooling would start, and George sputtered, having been caught unprepared.

"You should register," she told him. "Take out a loan if you have too. Life has a habit of getting in the way. If you want to do something, you should do it now."

Debbie was a student herself, working towards a degree in early childhood education. A natural with children, she wanted to teach preschool or kindergarten and dreamed about someday becoming a mother. The young couple soon discovered they had much in common, despite the opposite sides of the world on which they'd been raised.

Within two weeks of their first date, George found himself by her side in her parents' home, her four brothers also gathered around the table: their wives and Debbie's mother preparing dinner in the kitchen, Debbie's father asking George about his prospects. The Oslantski family was as warm and welcoming as George's had been cold and aloof. The home was loud with boisterous laughter and George hadn't felt so welcome anywhere since visiting his Aunt Ena's or the Mayo's. When Debbie reached under the table to hold his hand, he turned to smile at her and realized with a pang that he was home. Looking at her was like looking into his future. A real family. Finally.

It was never a question of if George and Debbie would marry, but rather, when. Within only a few months of seeing one another, George asked if they should start looking at houses. He'd saved up enough money to put a down payment on a nice, family home in Wainwright. Ace Construction was a fine job, and he felt he could put off flying school for a few more years.

"We could find somewhere near your parents," George suggested.

Debbie's face broke into a wide smile, but she shook her head.

"George," she said, "I can't live with you until you get that helicopter license. I won't be the reason why you don't achieve that."

Despite his hurry to start a life with Debbie, George was touched. He knew she was serious, so the couple was patient. To increase George's wage – making flight school a sooner possibility – George knew he ought to obtain a linesman's certificate, which would result in him receiving a raise in the industry he knew best, laying power line. Debbie helped him study and he was surprised by how quickly the ticket was achieved with her help. It seemed truly important to Debbie that George achieve his goals, just as she strove to achieve her own. Above all else, she trusted he wouldn't let her down, even after his success was achieved. George knew it was time to put the same amount of faith in her that she placed upon him.

He took her out for an evening drive, and they parked down a private dirt road.

"Debbie," he said, "There's something you need to know about me."

Debbie's face fell and George realized with dread that confessing his secret would be harder than he imagined.

"Oh god," she cried. "You're married!"

George roared with laughter and relief.

"Married? Oh god, no! I'm… I'm not supposed to be here, Debs. I'm illegal. I'm not supposed to work in Canada, and I came over on a three-month traveler's pass. My brother cashes in my cheques, and I pay taxes through a fake social insurance number."

Now it was Debbie's turn to laugh. She shook her head slowly, an incredulous look on her pretty face.

"Is that all?" She asked, still laughing. "I'm sorry, but that just doesn't seem very important. Like, who cares what's going on in little ol' Wainwright?"

George nodded, smiled, and kissed her. He felt better, and she was probably right. After all, she was the Canadian. She knew how her country worked, and it seemed many of the people he'd met in Canada were immigrants. Maybe his illegal status was a bigger problem in his head than in reality.

With Debbie now knowing George's long held secret, he started passing his pay cheques off to her instead of mailing them to Robert. Their bank account could only be in her name, but the young couple started saving for George's helicopter license and the home they'd one day share.

At Debbie's encouragement and with enough money in the bank after power lining for a few months, George finally enrolled in helicopter school. He got rehired at Ace Construction, where he could work while logging flight hours during seasonal shutdowns. Debbie and George studied together in his hotel room each evening, or in the front seat of his Lincoln, or in the den at her parent's house. To enroll in a condensed helicopter course, George first had to have two hundred hours of fixed wing flight under his belt, so he and Debbie made a habit of flying a two-seater Cessna 150 to Edmonton City Centre for coffee dates every Saturday morning. She was nervous in the plane at first, but forged courage. This was who Debbie was, George realized. He was wrong in thinking she was fearless. She wasn't immune to fear. She was simply *courageous.*

At that time in their life, it seemed as though every conversation revolved around flight school. They both knew his career in aviation was their ticket to an abundant life and future happiness. Debbie grilled him, quizzed him, reminded him of upcoming tests, and soon showed him she had probably learned more about aviation theory than even he had. She challenged him and wouldn't accept any comments about him being too tired to study after a long day at work.

"That's what coffee's for," she'd simply laugh.

With Debbie's help, George was performing at the top of his class for the first time in his academic career. He graduated with his helicopter license in December 1980, and was hired by Kenting Helicopters on the first of January 1981, one day after his twenty-fifth birthday.

George was thrilled to start at Kenting. While the company was part of a larger franchise, the owners of the Calgary base ran the outfit like a mom & pop shop, making sure their employees were well cared for and happy at work. George was set up in a small company house that Debbie could comfortably visit on her days off, and he returned to her in Wainwright on his.

Only a few days into his hire, George formed a fast friendship with another pilot who had graduated from his flight studies only a few years prior. Leroy Dean was a few years older than George, was already married, and was quick to assume a mentorship role. Their friendship was confined to work, at first, but George looked forward to seeing his new friend each morning. The men joked around like brothers, and Leroy was happy to fill any gaps in George's education or introduction to Canadian culture.

On one occasion, George was working a wildfire up north in the Fort McMurray district, near Birch Mountain. He was tasked with transporting firefighters and supplies to the fire site. While packing the helicopter, a technician threw a sleeping bag too high, hitting the rotor blade of the helicopter and damaging the craft so it wasn't safe to fly. George was livid, how would he get to the radio tower to call the problem in? Furthermore, how could he abandon such an expensive machine to replace its broken parts? Luckily, Leroy was quick to the rescue. He transported George to the tower and, bright and early the next day, Leroy returned with a new rotor head and set of blades, the doors of Leroy's helicopter removed to fit the large pieces of equipment. Together, they mended the broken machine. Leroy had taken George 'under his wing,' and, with Leroy's help, George picked up as many flying hours as he could, hoping greater experience would make him a more attractive candidate for future jobs.

But logging helicopter flight hours had become a feat in the years since George first considered Canadian aviation schools. Competition in the field was ramping up with every new pilot needing to gain those elusive hours, while every veteran pilot competed to fill them for high pay. But George tried not to rush life; he was too busy counting himself lucky for his love, and for the real, sustainable income he was making, and for the freedom of his adopted nation. For the first time ever, the future seemed like something he could navigate, like a road already paved. His would be spent with Debbie. Every dollar he earned would go towards the house he would buy her, and the children she'd someday give him.

But they say if you want to make God laugh, tell him your plans.

George had been eying up engagement rings the same week Kenting called him down, out of the sky. His employer had been applying for a grant to fund a mountain flying certificate for George, but there was trouble with the application.

George landed immediately; stomach knotted in dread. He called Doug McArthur, his chief pilot. On the other end of the line, Doug was cold.

"Your social insurance number is faulty," he said.

George felt his stomach drop. "Oh, uh… I have to get it verified," he lied. "Let me make a few calls."

"We need you to return to the base immediately."

George's blood froze. He hung up the phone and walked with his head in his hands. What was he going to do? He could give them another number, he reasoned. He could try his luck one more time. It might work. It had worked before. He got into his machine and flew back to the Kenting hanger, thinking of excuses all the way. Doug met him on the tarmac.

"So, the illegal alien has landed," he said.

The jig was up.

The Canadian government gave George two options: he could buy his own plane ticket, or the British Embassy would buy him one and he'd be indebted to them until the cost was paid back. George tried to reason with the government. He had paid all his taxes, after all, and he planned to marry a Canadian. He also had a good job in an industry that lacked qualified workers.

Immigration didn't care. George had to leave.

Debbie and her family were outraged, so much so that her father got involved with the situation on political levels. It didn't matter. Nothing helped. George was going back.

But not back home.

George didn't have a home. Rhodesia wasn't even Rhodesia, anymore. On the first of April 1980, the nation was renamed

Zimbabwe, and was governed by a communist leader without tolerance for anyone who had previously pledged allegiance to earlier governance.

For George, there was nowhere to go back to, there was only *go.*

He signed everything he had over to Debbie. He put his Lincoln in her name, his dirt bike too. The savings they'd amassed stayed in her name, and he took just enough cash out to get himself to South Africa where he'd stay with Clive Mayo until things got sorted.

"I'm coming back," he promised Debbie as they drove the Lincoln to Calgary Airport.

Debbie nodded and tried to smile through her tears. "I know," she said, reaching for his hand.

They listened to country music on the radio, which George didn't typically enjoy, but Debbie did, having grown up on the sound. It was fitting music for two broken hearts.

At the terminal, Debbie told George she loved him, and he told her the same. They kissed one last time before there would be an ocean between them, and then two immigration agents gestured that it was time for George to go.

He surrendered his British passport to the officials, and they walked him through the airport, through customs, and stayed with him until he was deposited in his seat on the airplane before handing his documents back into his possession. Other passengers watched with wide eyes as George was sat, escorted as though he were a dangerous criminal. George kept his eyes down. He had known disappointment, but he had never known a blow so crushing, or a scene so embarrassing.

When the plane took off, George couldn't bear to look out the window as the only place he'd ever truly felt was home disappeared from view.

SOUTH AFRICA, AGAIN

September 1981

George could barely bring himself to stand and disembark as the plane touched down on the African runway. George's trip had required a transfer in London, UK, before arriving in South Africa, and now British passengers pulled their belongings down from overhead cabinets, their bodies bumping into one another as they vied to exit the craft, stripping themselves of jackets and hanging too-heavy clothing over their arms in anticipation of the heat they'd heard about. George just waited and looked out the window, feeling the eyes of strangers on him as they walked by.

Leaving the plane to receive his luggage from the baggage carousel, George felt about the loneliest creature on earth. It was as though he were the last of his kind – *The Last Rhodie* – he thought, a beast doomed to walk the earth in solitude, forever denied a place to make his home. But when he entered the main part of the airport, he saw that the Mayos were there: Mr. and Mrs. Mayo, Neil, Clive, and Clive's wife, Carole. They stood there with sympathetic smiles on their faces, happy to see their boy, sad to see a fellow Rhodie kicked down for the umpteenth time. Mrs. Mayo was the first to come forward, her arms outstretched.

"Oh Love, we're so sorry you've been sent back, but it's so good to see you."

George closed his eyes. His throat felt tight. Hugging Mrs. Mayo reminded him of his own mother, and Mrs. Mayo seemed to sense it. She squeezed him a little tighter, patting his back.

George wasn't alone.

"So, Canada cottoned on to you, son," Mr. Mayo teased, reaching to shake George's hand as his wife broke the embrace. Mrs. Mayo pretended to cuff her husband's shoulder, shushing him.

"But not before you made off with one of their most precious natural resources, eh, George?" she winked, referring to Debbie. "Now then, let's get you home and make plans to get that girl of yours home, too."

The group retreated to Mr. and Mrs. Mayo's home in Johannesburg. The Mayos had emigrated to South Africa as the Bush War spilled into small towns, shortly after George had left. George was to stay at Clive and Carole's home, nearby Mr. and Mrs. Mayo's. Neil didn't have a place set up with enough room yet, but George was grateful to be in the warm, relaxed company of old friends.

The first thing George did when he reached the Mayo home was call Debbie, who sounded glad to hear from him although sadness was still obvious in her voice.

"It'll be okay, Debs," he tried to tell her. Heartened by friendship, he now believed it to be true. "This isn't the end. I'm coming back. I promise."

The next day, he went to work, already making his plans a reality.

First, George applied for a work visa in South Africa. As a helicopter pilot, he was almost immediately granted landed immigrant status.

Coincidentally, Clive had established a career as a helicopter mechanic and had a good job at a nearby airport. Even luckier, pilots in South Africa were in high demand, so it was easy for George to cross off the first task in his plan to achieve status in Canada. He secured a job and got to work completing the thousand hours of flight time necessary for complete Canadian qualifications.

Generally, George's job was to fly power company workers – laborers and linesmen, as he had once been – from their rural communities to their job camps. It was easy, steady work, and George was glad of it. He had plenty of leeway to use the machines as he liked, too. Knowing Seb had set down roots in South Africa after leaving London, he surprised his old friend by landing on a tiny patch of grass in Seb's backyard to visit while he had friends over for a party.

"God, you sure have a way of showing up, don't you?" Seb had joked. "Hey mate," he said, gently, "Have you heard? Eva O'Kelly has married that Lieutenant Warren."

"Who?" George asked as he jumped out of the helicopter.

"Come on," Seb said, raising his brows at George's reaction. "You know who."

George just shrugged and smirked. Having found a love so much truer in Canada, he couldn't have cared less.

George spent every hour he could in his craft. Every hour he logged made him a more attractive candidate for Canadian citizenship, and every hour brought him closer to Debs.

The couple spoke over the phone every day, racking up enormous telephone bills speaking so often by long distance. On her side of the world, Debbie worked a good job at a daycare center, having graduated with a diploma in Early Childhood Development. Calm and nurturing, she made the perfect nursery schoolteacher. Because of the time difference, George was forced to often call her at the end of his day, which was the middle of hers. She was always glad to hear his voice, despite having to keep conversations short to care for her young pupils.

One day, George called with something more than a simple hello.

"Debbie," he said, hearing the little children play in happy, singsong voices in the background. "Debbie, would you come to South Africa and marry me?"

Debbie started laughing before she could reply in words. "Yes! Of course, I will, George! When?"

"As soon as you can!" George blurted out, right back. "I know it won't be a big wedding, and I know your parents probably won't be able to make it…"

"It'll be perfect, George," Debbie promised. George sighed into the phone, knowing if she said something would be so, then so it would be.

Debbie had a lot to do before leaving for South Africa. She'd never travelled outside of Canada before and was nervous about the ordeal. She'd need multiple vaccinations to visit the continent and could barely comprehend how she'd move through customs at three different international airports before arriving in Johannesburg. But Debbie, as George knew her to be, was courageous. In a matter of weeks, she'd made the trip and, at last,

George held his fiancé in his arms.

South Africa, in response to how Rhodesia had failed under segregation, had softened its apartheid laws since the establishment of Zimbabwe. Already, it was a different country from the one George had known in the seventies. Now, Black Africans were welcome to visit any restaurant they preferred, and frequented movie theatres and many other parts of the communities which had once been reserved only for whites. They were achieving greater positions of power in the workforce. But while racial progress was a boon for a more equitable society, government corruption was still obvious in the imprisonment of civil rights activists such as Nelson Mandela, and so South Africa's social and political climate was tense and dangerous. In the mere months since George had arrived, he had gone from feeling complete safety in the country, to refraining from walking after dark.

The feeling of unease was one of the first things Debbie noticed when she landed in early December.

"They even keep the windows locked!" Debbie whispered to George after Mrs. Mayo gave her a tour of the house.

"Just at night," George laughed.

"My parents don't even lock the front door!" Debbie exclaimed. "I don't think I'm in Canada anymore," she shook her head, bewildered.

But Debbie loved Africa. Her fascination was endless: she was interested in even such little things as the different texture of grass, telling George it simply wasn't as soft as in Canada. She loved that Africa's soil was a rich red and had so many questions about the pigment. Was it red because it was iron rich, like they said soil was in Prince Edward Island, where it was ideal for growing potatoes?

To George, the continent on which he had grown up seemed new when exploring it in Debbie's company. He bought a secondhand mini cooper in anticipation of her arrival, and together they ripped around backroads. Once, the couple was caught in a flash flood.

"George! The backseat is all water!" Debbie shouted, her body turned to peer into the back, mouth agape.

"There's a hole in the floorboard," George explained, "We just gotta keep driving until it stops raining!"

"Is this vehicle a car or a boat?" Debbie asked, nearly

hysterical with laughter.

They drove with a swimming pool in the back of the car until the rain stopped. Then, within the hour, just as George said it would, the water rushed back out of the hole in the backseat floorboard, as though someone had pulled a bathtub plug. Within moments under the hot sun, the vehicle was once again completely dry.

"What a place this is!" Debbie shook her head in wonder. "What a place."

Debbie, Carole, and Mrs. Mayo spent the weeks leading up to Christmas preparing for the wedding that would be held on January 8th, 1982. Mrs. Mayo called Debbie her new daughter and thought she was the best thing since sliced bread. She waited on Deb hand and foot, even though Debbie needed no extra attention and simply wanted to be helpful to her gracious hosts. One day, soon after Debbie had arrived, Mrs. Mayo took her honorary daughter's face in her hands and inspected her ears.

"My girl," Mrs. Mayo said, "if you go get your ears pierced today, I'll have a present for you when you return."

Debbie was ecstatic. Almost twenty-five years old, she'd never had a reason to pierce her ears before. Carole took Debbie into town to a salon where they performed the service for a small fee.

When the girls returned to the Mayo home, Mrs. Mayo was waiting with a small box dressed up in a bow, just for Debbie. Debbie opened it to find diamond studs inside.

"I can't!" Debbie gasped. "They are too beautiful!"

Mrs. Mayo tut-tutted Debbie, shaking her head and smiling. Opposite the women, Carole smiled, too.

"I've given all of my daughters jewelry as they came into our family," Mrs. Mayo said.

Carole smiled and nodded. "It's true!"

"And this isn't the last you'll get from me," Mrs. Mayo promised, helping Debbie swap the salon's cheap baubles for her new studs. "You must know, Debbie, George is very, very important to us. He's our family, and we are so grateful he has a person as wonderful as you in his life. And now you are our family, too."

Debbie choked back tears and hugged her adopted mother-in-law.

But a few days later, it was Mrs. Mayo's turn to cry, and not in joy. George was scheduled to spend a few days right before

Christmas flying in Botswana. Knowing that a rift had developed between George and his father, Debbie made plans to travel to Zimbabwe. Treasuring her own parents so, Debbie simply couldn't accept that Mr. Kelham wouldn't attend his youngest son's wedding. She wanted to meet the man who would become her father-in-law.

George told her it wasn't a good idea, but Debs had insisted. Reluctantly, George bought Debbie a plane ticket, booked her a hotel in Harare – the city that had once been known as Salisbury – gave her his father's address and phone number, and wrote detailed instructions on how to take a taxicab to a hotel.

On the morning Debbie was meant to take her short trip into the African interior, Mrs. Mayo knocked on the door of Debbie's bedroom as she was getting ready to leave. When Debbie answered, she was surprised to find the usually cheery lady in tears.

"Please don't go, Debbie," Mrs. Mayo begged. "You are too sweet, child. I don't want you to get hurt."

Debbie was touched, but she wasn't scared. She'd been on the phone with John Leonard only a couple days earlier. He knew she was coming, and she was excited to meet him.

"Mrs. Mayo," Debbie said, "you can't truly believe George would send me somewhere dangerous! I'll be very safe."

Mrs. Mayo shook her head, her tears running down her face.

"Oh, lovely girl," she whispered, "that's not what I'm worried about."

Not knowing what else to do, Debbie embraced Mrs. Mayo and promised her that she'd be careful.

Later though, Debbie would reflect upon that promise and realize she hadn't totally been sure what "careful" meant. It wasn't until she'd landed in the Zimbabwe airport and walked onto the outdoor tarmac that a sudden and terrible thought filled Debbie's mind:

I don't know anybody here, and George is in Botswana.

For the first time in Debbie's life, she was truly alone.

But she followed George's instructions to a tee. She flagged down a cab and asked the driver to take her to the hotel where George had made the booking. The driver was friendly and polite, and sat beside another man in the passenger seat, who kept him company in the front.

As the sights of Harare flashed by outside the window of the cab, the interior of the vehicle filled with the musical sounds of Swahili, a language that Debbie, of course, was completely unfamiliar with. Still rattled with the truth of the fact that she was completely out of her depth, a chill fell over Debbie.

How do I know where they are taking me? She thought. They could be taking her anywhere at all. The men could take her right to their home, if they so chose, and what could she do about it? No one knew where she was, and she didn't even know how to call the police. In fact, she didn't even know if Zimbabwe had a functioning police force!

Completely alone.

But the men were good and honest. Within minutes of the drive, a hotel with a name she recognized from George's instructions came into view. Relieved and ready to hide away in her hotel room, Debbie handsomely tipped the driver and his companion for her safe delivery.

She checked in and unpacked what few things she had brought with her to Zimbabwe, and then called John Leonard. She couldn't wait to take him out to dinner and hoped that he might already be in town so they could meet that night as well as the next, as planned.

The phone rang and rang. Debbie was about to hang up when, finally, an irritated voice answered the phone.

"Hello, Mr. Kelham!" Debbie said brightly. "It's Debs! I just wanted to let you know that I landed safely and that I'm here in Harare in a hotel now. I'm really looking forward to meeting you tomorrow! Are you in Harare, or still in Umtali – um, Mutare, I mean?"

There was a pause on the line and for a moment Debbie worried they had lost the connection.

"No," John Leonard finally answered. "No, I can't make it into Salisbury."

"Um… no?" Debbie asked. She couldn't believe what she was hearing. "Mr. Kelham, I would really like to take you for dinner," she said, not knowing what else to say.

"You could give Marion a call," he said. "She's in the city. She might be available. Hold on, I'll get you the number."

Debbie's eyes so stung with tears that she could barely see the notepad in front of her as she scratched down her soon to be sister in law's phone number. Before Debbie could say goodbye, John Leonard had hung up the phone.

Debbie took a deep breath. She steadied herself and, insisting that the trip not be for naught, she dialed Marion's number, inviting George's sister to meet with her over lunch the next day.

"I'll have to bring my boys," Marion said, coolly.

"That's fine," Debbie said. "I'd love to meet my nephews!"

The next day at noon, Debbie waited at a restaurant.

Upon first sight, Debbie was shocked by Marion's bright, beautiful red hair. One of the little boys had a similarly colored hue, and Debbie was almost going to make a joke about how good it was to meet her and see this family gene in action, and how she wondered if she and George would have a child with such gorgeous hair. But before Debbie could say anything – in fact, while Marion was still seating herself and the toddlers - Marion asked Debbie a question point-blank.

"Are you Catholic, then?"

Debbie was taken aback.

"No," she reported. "I'm not really anything. I suppose my family would be considered Protestant, but we've never really been regular church-goers."

Marion was stone-faced. She lifted a brow.

"So, you'll convert? After or before the wedding? Our mother was protestant before marrying dad, but she converted before they wed."

Fat lot of good religion did him, Debbie thought, still sore that her father-in-law wasn't interested enough in his son's life to meet the woman he was about to marry.

"No." Debbie said. "George isn't religious. I don't think it matters to him. I don't plan on converting."

Marion shook her head. She looked Debbie up and down.

"It will," she said. "It will matter a great deal when you have children. Can you at least tell me you'll raise George's children in the church?"

Debbie felt her heart drop. There was no saving this relationship with her sister-in-law, there was no even *forming* a relationship with George's family. Suddenly, she felt very tired and very heartbroken for George. She wished he was closer. She wanted to touch him and promise his future with her would know nothing but unconditional love. Debbie looked Marion straight in the eyes and shook her head.

The young women ate an awkward, almost silent, and very quick lunch before Marion buckled her children back into the double stroller and left the restaurant.

Debbie's face was still swollen from tears when Mr. and Mrs. Mayo picked her up from the airport. Mrs. Mayo began to cry at the sight of Debbie's unhappiness. She hugged Debbie but waited until they were in the privacy of their home to ask what had happened. Debbie's strong, optimistic exterior broke down in the presence of someone so kind.

"He didn't even meet me," Debbie cried. "And George's sister – that Marion – she doesn't want anything to do with me, just because I'm not Catholic!"

Mrs. Mayo shook her head. "Oh Love, I knew you were too good for the likes of them, Debs. Your George is the best to ever come out of that house. This is why I didn't want you to go. I knew they'd be mean. They're mean, hard, cold people. I'm so sorry."

Hearing part of the conversation from the hall, Carole came into the room.

"What's going on?"

Mrs. Mayo gave her a look. "Didn't go as planned," she said, and Debbie realized the Mayo women had spoken about this possibility between themselves.

"Oh, Debs," Carole frowned. "Oh, that's terrible."

"But you're *ours* now, lovely girl." Mrs. Mayo insisted.

Carole nodded. "That's right, Debs. We're your family, now. We're the in-laws. And there's no escape!"

The women laughed and cried a little bit more, all over again. Debbie felt better, strengthened by the friendship she'd found in the Mayos, and fortified by the promise she made herself, that George would never again know love except that which was warm, welcoming, and unendingly his.

~

A couple days later, George returned from Botswana and the young couple were ready to be wed. Christmas was only a few days away, but few around them seemed to even notice the event for excitement over wedding planning.

The couple planned to be married in a courthouse but had accidently left it too late, not knowing they had to make an appointment weeks prior. They had assumed it would be like the movies - they'd just show up and do the deed!

Disappointed to learn that they would have to wait, a helpful courthouse clerk suggested they call a Canadian minister who was visiting the country and who had legal status to perform marriage ceremonies. Debbie and George were doubtful. Certainly, she'd had enough to do with religious rejection to last more than just the trip, and they didn't know if a protestant minister would bless the union of a lapsed Catholic and a non-churchgoer. But low and behold, the man was happy indeed to sanctify the future of the young Kelham family… just as long as it was on Wednesday at seven PM sharp, because that was his only free day of the week.

The couple married in a lovely old church with a clock face on its steeple, but no clock hands to track the time. Debbie thought that was a good omen, as they were promising each other forever, time beyond time. She wore a white dress she'd bought in Canada at Sears for $29.99, and a hat she'd found in South Africa. Mrs. Mayo put together a bouquet of roses so beautiful that Debbie protested that they ought not to be cut from her garden, but Mrs. Mayo wasn't having it. She lovingly pieced together the bouquet, as well as boutonnieres for the men, too.

Watching his bride joke and laugh with the mother of his best friends, George thought he was the luckiest man on earth. If he was not to have come from a place of complete love, then he was exceedingly fortunate to have found it.

All in all, eighteen people attended the wedding, including the Mayos and Sebastian Wince. But many more loved the couple and celebrated their union by sending gifts and prayers and well wishes – Debbie's dear family included. George and Debbie treated their guests to a first-class meal with drinks, all for only three-hundred-dollars, and then George went back to work the next day.

But this time, Debs went with him.

~

George's newest gig was providing transport in and out of Sun City, an exclusive resort built on traditional tribal lands, which meant the establishment could offer the alcohol, gambling, and shows that puritanical South Africa had outlawed. Most of the pilots on George's team didn't really like the work as it meant staying at the club for long weeks away from home and family, but for George and Debbie, being put up in a snazzy hotel was the perfect opportunity to honeymoon.

Prior to the wedding, George got his employer's approval to bring his new bride with him on the assignment.

"That would be fine, Mr. Kelham," his employer nodded. "So long as you promise to get enough sleep," the man had winked.

Debbie was to relax by the pool while George flew during daylight hours, then they'd enjoy the country club's fine restaurants and their beautiful bedroom by night. On the first morning George left Debbie to enjoy herself, he reminded her to apply a lot of sunscreen while she was outside.

"Oh, don't you worry about me, George!" Debbie said, irritated at having been told what to do. "The sun is the sun! It doesn't matter where you are. It gets hot in Canada, too!"

George decided he would say no more on the subject and thought that the interaction had probably served as his first lesson on marriage. But when he returned to the hotel that night, he found her laid up in bed with big blisters all over her scarlet legs.

"I'm burned," she wept. George felt awful, but he couldn't help but laugh. Debbie laughed too, through her tears. They had both learned their first lesson on marriage that day.

After a week and a half of wearing long skirts, Debbie recovered enough to enjoy her time at the country club, and soon it seemed the couple experienced an exciting new story each day. It was a good time for Debbie to visit, as the resort was hosting a million-dollar Sun City Golf Classic. Each professional golfer had been sponsored by various celebrities, who had accompanied them to the resort in entourage. The competition itself was dazzling, with prizes on each hole. The professional golfers could drive away with a new Lamborghini if they hit a hole in one on hole sixteen (alas, no one did!).

The Golf Classic brought a crowd of celebrities into Sun City. During their star-studded honeymoon, George and Debbie ate at a restaurant table beside actor Glen Campbell, and, another time, beside the American actor, Telly Savalas – the famed television show detective. George was thrilled to spend a day taxiing Sean Connery and his wife to and from a local game reserve while they too visited the resort. Later, Debbie and George watched The Beach Boys play live in concert and once they stepped into an elevator to find James Hunt already inside. The English Formula One Champion was unaccompanied by any sort of security and was friendly, chatting with George and Debbie as they waited for the elevator to reach their desired floors. George couldn't believe they'd stumbled into James Hunt's company. George enjoyed watching Formula One races and was fascinated by the fierce competition between Hunt and his rival, Nikki Lauder, a story which would later be immortalized by the movie *Rush*. Lo and behold, only a few months later, in February, George would have the pleasure of meeting Nikki Lauder, as well, as George was assigned to fly Lauder to Kyalami Grand Prix Circuit for a Formula One event.

Going up!

But despite all their fun, by the end of January, it was time for Debbie to return to work. The couple shared another tearful parting, although it was with much less sorrow. Mostly, Debbie cried as she hugged the South African crowd goodbye because she knew she'd miss her new friends so much.

Now that the couple was legally wed, George had the right to reside in Canada with his new wife. It would just take a little bit of paperwork. George was sure the filing would be completed in a week's time, but in the end, it took three more months before George was let into Canada. The Canadian Embassy was suspicious, George knew, and rightly so, but the Kelhams were truthful when they claimed to have loved one another long before George's deportation.

They would love one another long after, too.

PART THREE

CALGARY
March 1982

Debbie was there to meet George when he arrived at the Calgary airport, and it was almost as if he'd never been gone. With a thousand hours of flight time under his belt, landed immigrant status, and a work visa, George was hired back on at Kenting Helicopters, and the newlyweds purchased their first home – the title held in *both* their names.

The purchase was a little mobile home in Calgary and their mortgage had an interest rate of twenty-one percent – 18% of which was financed through the bank, with the remaining 3% funded through the government. Debbie got to work right away filling the space with creature comforts from secondhand stores and garage sales. She'd inherited a three-legged couch from a brother, and George stacked bricks under the missing leg to balance the thing. The set up was modest but homey, and, best of all, it was theirs. George had never lived with a woman other than his mother before, and he took great pride and pleasure waking up with his smart, beautiful, capable wife each morning, devouring her delicious home-cooked meals every evening after work.

Initially, George had worried there would be some bad feeling among the crew at Kenting when he returned to work after the way he was forced to leave, but on George's first day back, Leroy was the first to meet him. All others welcomed him back to the team with gusto, and George was grateful to be accepted into the social fold. But not everyone was there. The crew had significantly shrunk in size since George had left.

"Where is everyone?" George asked Leroy.

"You're playing a game of last man standing," Leroy joked. Although George had completed his training with fourteen other men, only nine had finished their course. Four of the missing pilots had given up flying almost immediately after earning their licenses; one because he had been in an accident on his first job. The fifth had perished a plane crash.

George shook his head in news of the tragedy but couldn't believe there were men who willingly gave up the job. George had worked his whole life to get his helicopter license. He couldn't fathom ever walking away – not when instead he could fly.

After a couple uneventful weeks of work, immigration showed up again at Kenting Helicopters.

"Mr. Kelham," one of the officers started.

"Yeah, yeah," George said. "You're here to make sure my marriage isn't a sham."

The immigration official stammered.

"It's fine," George interrupted. "I'd be a fool to think you wouldn't check up on us. Here." George handed the officials a copy of the deed to his home with Debbie's name listed on the document alongside his. "We have a home, I'm back in my old job – the same one you guys made me give up when I was deported – and I'm paying my taxes. Same as I always was. I'm not a criminal, and I'm not going anywhere."

The immigration officials were convinced. They didn't visit Kenting Helicopter again, but sadly, George was wrong on the last count.

By the end of 1982, Alberta was facing recession. Calgary, a city with an economy that hinged on the price of oil, suffered more than anywhere else. In anticipation of upcoming layoffs, George and Debbie sold their mobile home and moved to the cheaper community of Wainwright, where Debbie was hired back on to her old job at the daycare center. They were right to do so. Only months later, Kenting Helicopters laid off five men, George included.

George spent weeks pounding the pavement, looking to accept any work available. Debbie's work at the daycare was safe, but her pay wasn't nearly enough to cover the couple's shared bills. Things were looking precarious, and, for the first time in a long time, George started to feel nervous. He'd faced poverty before, but it had never mattered. He could miss meals, he could share a flat (or even a bed) with a friend, and he could make do without. But now he had Debbie, and while she asked for nothing, he wanted to give her the world. He needed to provide for her. She deserved a good home, food on the table, and anything else she wanted.

The Kelhams were just about broke. George was sure he'd applied for every open position in the province and was contacting everyone he'd ever known in any professional capacity when, finally, a contact was able to help. Andy Veal, or "Porkchop" as he was known at Ace Construction, confided that he knew where George could work, but he'd have to be comfortable accepting cash under the table. After all, Porkchop informed, the job was in America and, technically, George would be an illegal immigrant.

George stifled a groan. It seemed his past was inescapable, but he couldn't turn work down. On Porkchop's advice, George drove to Denver, Colorado, and was hired on to build powerline for cash in the cities of Grand Junction and Fruita.

The uneventful job lasted six months. George and Debbie missed one another, separated again, but they knew love was patient. Whenever possible, Debbie visited George in Denver, just as she had when George was exiled to South Africa. George kept his head down, worked hard, and when the job came to an end, George was thrilled when Andy Veal sent yet another power lining project his way – this time, a lot closer to home.

The gig was in Beaverlodge, Alberta. George was put up in a motel, made regular money, and got to go home on days off – a great deal more than many Albertans were blessed with at the time. Still, George grieved the loss of his career in aviation. He had worked so hard to become a pilot and overnight he had found himself grounded. He had applied to work with so many helicopter companies that he'd lost count, but no one was hiring. Most days, George tried not to think about the loss, otherwise it took his breath away.

But not thinking about the sky and the machines was next to impossible. Every day on his morning commute, George passed the Grande Prairie airport, and it was like salt rubbed into a wound. Every time, George was filled with a sense of longing. He longed to stand on the tarmac and enter the large hangers. He wanted to be near the beautiful machines, and take them to the sky, and again feel that sense of mastering the impossible, willing the iron cage around him to hover hundreds of feet above the earth.

Finally, one hot day in August, George couldn't take the heartbreak any longer.

George was on his way to Beaverlodge when he pulled into the parking lot of the airport instead of continuing down the road to the powerline site. He turned with such conviction he surprised himself, feeling it was almost as if someone else had control of the wheel. George walked into the hanger and met a man named Henry Stevens[5], who introduced himself as the operations manager of Harmony Helicopters[6].

"Henry," George said, after introducing himself as a pilot. "I will sweep your floors, wash your helicopters, and fly too, if you want me too. I need to be around these machines again."

Henry's eyes narrowed. George had blurted his proposal while the men were still shaking hands. Slowly, a knowing smile came over Henry's face.

"There's a fire south of High Prairie, east of us. You know the area?"

George nodded. "Yes, sir."

"Okay, I'll put you on it. Let's see how you do."

Just like that, George's luck had changed, but he also knew luck wouldn't sustain success as well as hard work would. To show his appreciation for the job, George worked for Harmony Helicopters long past the hours he was paid. He got to the hanger early and stayed long after all machines were grounded for the evening. He swept floors and washed crafts he hadn't even flown.

Comfortable in the vicinity of another man's hard work, Henry kept George company from the seat of a lawn chair he had dragged into the hanger. The man drank like a fish and smoked like a chimney and George was constantly sweeping cigarette ash from the floor. But it didn't matter to George what Henry did or didn't do at work. All George cared about was the fact that he, once again, had a fulltime, stable, and secure position flying the Canadian sky. He couldn't imagine asking life for even one more blessing. Then one came, anyway.

Sweet Debbie was pregnant with the couple's first child.

---

[5] Name fictionalized.

[6] Company name fictionalized.

George was ecstatic.

"Are you hoping for a boy or a girl?" Debbie asked him, after sharing the news.

George just smiled and shook his head. "As long as you are okay, and our child is okay, Debs. That's all that matters."

By the end of George's first summer with Harmony Helicopters, Debbie was beginning the second trimester of her pregnancy and things were progressing perfectly. George couldn't stop looking at his wife; she glowed as a mother-to-be and he admired how gracefully she adapted to best nurture their unborn child. It seemed her maternal instincts were strong right from the start. She seemed to be constantly thinking about the child she carried, and she prioritized the pregnancy by eating nutritious food, limiting all stress, and taking gentle walks. George could only hope he'd rise to the responsibilities of fatherhood with the same mastery she possessed as she became a mother.

It had been a busy summer for George. After spending months fighting wildfire from the sky, George received a telephone call from Henry while he was stationed in High Prairie.

"Are you studying for the A-Star test?" Henry wanted to know.

George was surprised by the question. "Why do you ask?"

"Why do you think?" Henry laughed. "I'll tell you what's going to happen. We're going to pay for your A-Star endorsement, and then we're sending you and a machine to Lesser Slave Lake. We'll give you a pick-up truck and an apartment. The Lesser Slave base is unmanned, but Harmony Helicopters wants to prioritize the area. They want you to be base manager."

George could hardly believe the offer. The idea of managing his own helicopter base set him on fire: his breath quickened, and heart leapt, and a shiver of excitement passed over his body. Still, he suppressed his delight and told Henry he'd need a couple days to think it over. There was Debbie, after all, and soon, a new baby. George wasn't sure he could ask so much of his wife – to move away from family, again, right as they were forming their own. That night, he tried to remain neutral in tone as he told her what new opportunity had presented itself. If she truly didn't want George to go, he wouldn't pressure her to grant approval.

Of course, Debbie saw right through him.

"George," she said, "This is a really big deal!"

George dropped the act and leaned in closer, relieved he was able to show his eagerness.

"So, yes? I should take the job?"

Debbie laughed. "Yes, George. Yes, you should be Slave Lake's base manager!"

A smile spread over George's face. Life was about to change all over again.

SLAVE LAKE

September 1st, 1983

Until George was inside the A-Star and on route to the little two-horse town he'd been promised a managerial position in, he didn't truly believe it would happen. It felt like only a moment ago he was laying line, bottom of the totem pole in employment status and getting paid under the table while he dreamed of the sky. But Henry was true to his word, and George was just about the happiest he'd ever been while flying over Lesser Slave Lake.

He had been told Slave Lake was tiny; the kind of place that had had a dirt road for a main street only a few years prior, and that that street was so short a person could stand on one end and spit to the other. From the sky, he could see that only part of the rumor was true: the town was small, with perhaps three dozen shop buildings in the business area, but its main street wasn't short. Slave Lake had made use of the wide expanse of land it was set on and while it wasn't sprawling, the town wasn't nestled in the woods, either. It had room to grow free, wide-open to possibility. Flying overhead, George had the sense that the town, though still humbly emerging, would build itself with confidence – following a similar process that George, himself, was following.

The landscape below him was beyond stunning. George was amazed to see the enormous, glistening lake was shored with white sandy beaches instead of river rock. It was as if someone had somehow plucked the best part of South Africa and dropped it right into the center of Canada's boreal forest – the view from the sky a high contrast palette of sun-kissed tan, deep greens, and bright blues.

Following instructions, George landed the A Star next to a hanger in the municipality's industrial area. Water bomber airplanes sat parked on the tarmac. George had been told that there would be a truck waiting for him outside the hanger, and that all he had to do was go into the office area and find the weatherman, who would have possession of the keys to both the vehicle and George's new apartment.

George entered the building feeling a little unsure. He couldn't remember the name of the weatherman and didn't know what he looked like. He slowly entered the hanger.

"Well, hullo!" a voice said.

George turned, his eyes still adjusting to the dark after flying through sun. A man approached him with an outstretched hand.

"Gerry Labrie," the man said as George shook his hand. "I'm a fixed wing pilot here with Peace Air. You new to town?"

George nodded and introduced himself. He told Gerry he was looking for a weatherman.

Gerry helped George find the weatherman and then introduced him to others in the hanger before giving George directions to the apartment Harmony Helicopters had rented for George in town.

"What are you doing tonight, George?" Gerry asked before George drove away. "A bunch of us meet for a bite to eat at Carlos' Restaurant after work. Come by and I'll give you the run down on Slave Lake."

George accepted the offer. He was grateful to have found a friend so soon and was eager to learn the lay of the land and to get to know its residents.

The apartment Harmony Helicopters had rented for George was in the south-east quadrant of town, located behind a small convenience store the locals called Tags. There were two bedrooms in the apartment and George planned to buy a small, second bed for the empty bedroom, imagining with much happiness how his little son or daughter would visit him when Debbie came by on weekends until they found a nearby home to call their own. For as fortunate as he felt for the new career opportunity, George also felt a pang of homesickness thinking about Debbie. They'd decided she would stay in Wainwright, near her mother, for at least the rest of her pregnancy. George was wary of putting his wife through such a big move when she already had so much on her plate and there was no guarantee the Slave Lake project would work out. It was probably best Debbie got to know the town in the same slow but safe way a person might gingerly step into a pool from the shallow end.

But George was going to cannon ball in.

The first thing he did after surveying the apartment was call Debbie from the phone Harmony Helicopters had already set up in his new living quarters. He told her he was settled and that he loved her, and then he unpacked his small bag of belongings. He locked up the apartment and took himself for a drive through town, circling streets until he found the place's only grocery store, and then he went inside to buy food to fill his empty fridge. When these errands were done, George found his way to Carlos' Restaurant and met Gerry Labrie in the folksy café.

~

George had given his all in every position he'd ever held, working as though his employment hinged on how much money he could save the company or make the boss, which was often true. But George had never worked as hard as when he became the Slave Lake Base Manager for Harmony Helicopters.

He knew his position at the base was a trial. If the Slave Lake outfit didn't bring in enough clients to keep it going, George would find himself back in Grande Prairie working under Henry Stevens again for reduced pay, no longer a manager of anything. George couldn't have that. Slave Lake was too new, too beautiful, and had too much potential and, finally, George was in a position of leadership. He knew he needed to succeed. He needed to bring Debbie here. In his heart, this was where they were meant to raise their family.

Without nearby family or friends making demands on his time, George worked night and day. He threw himself into building the base, making strategic plans on how to optimize operational efficiency. He pounded the pavement and kept his eyes on his pager, and when he wasn't meeting people on location or at their place of business, he had an office phone glued to his ear, cold calling new clients, setting up new accounts with local industries. He inserted himself into the business community, forging professional relationships with men who could pad the pockets of

Harmony Helicopters, for no benefit to George except that he might remain in his new role. He needed to be as much a salesman as a pilot, George realized; he needed to pitch hard and be willing to fail often, to get back up immediately and shake the dust from his clothes.

So, he did. It began to work.

The recession that had slowed down Calgary was of little evidence in the north. No place was as busy as little old Slave Lake, Alberta – a town that begun to make headlines for hosting more self-made millionaires per capita than any other place in Canada. Industries employing hard men couldn't find enough workers to meet their need and had resorted to throwing money at anyone who could show up – even if they only stayed on site a day or two before abandoning their post to blow cash on liquor or coke or fast women. In workers, there was little loyalty to one's employer when another job was just around the corner.

But that was exactly what set George apart.

One of the first accounts George brought to Harmony Helicopters was Steady Construction[7], a company that was in the process of being bought by a man named Willy Light[8], who was the original owner's brother-in-law. Owner of one of the only construction companies in northern Alberta, Willy didn't have time to drive to the many small and distant communities to scope out projects. It was much more efficient to rent a helicopter to fly to sites and put a bid together.

George flew Light and the Steady Construction team all around the northern Slave Lake region to look at lease sites and build roads. Being in the construction industry, Light was used to being let down by contractors. Once he realized he could rely on George like clockwork, Light sang his praises from the rooftops. He started feeding George the names and contact information of higher-ups in the oilfield industry, the kind of decision-makers Light had connections to on account of his newfound wealth and status. George would fly Steady Construction all morning and make calls in the afternoon. Anytime he wasn't flying existing clients, George was meeting new ones. Within a few months, he'd locked down dozens of contracts, including casual hiring with Alberta Forestry.

---

[7] Company name fictionalized.

[8] Name fictionalized.

The head owners of Harmony Helicopters in Hynes Creek were impressed, but not Henry Stevens. Despite the geographical distance between them, Stevens was still George's higher up, and he ordered George around as though George was a teenager, liable to sleep in for work or simply not show up at all. The attitude got old fast. George hadn't been talked to with such disdain since he left the Rhodesian military, and he certainly didn't appreciate it with the kind of money he was bringing in for the company. George doubted Stevens had ever booked so much work, and George hadn't so much as taken a sick day since his hire. Finally, in March 1984, Henry Steven's disrespect could no longer be tolerated.

George was in the air thinking of the new clients he would call later that day when he received a page from Beacon Answering Service. When he landed, he called the service operator to see what message she had to relay.

"Pilot Kelham," the voice reported over static, "Please call Debbie."

George's heart soared. Debbie wouldn't have left a message unless it was important. Would his child enter the world today? He hung up the payphone and called his wife at home, cheering when she told him, yes, the baby was coming, and that he needed to meet her at the hospital. Next, George called Henry Stevens.

"Henry," George said. "I'm sorry to do this at last minute, but my wife is in the hospital having a baby and I need to leave for Wainwright immediately."

Henry just laughed. "Like hell you are! Who's going to finish your job?"

"We have contract pilots," George said. He was stunned his supervisor would treat him so poorly, especially considering how rare it was that George asked for anything at all. "Send someone to cover me, just for a day or two. I have to get to the hospital."

"You've got until tomorrow morning," Henry said. "You've got a week left of work and a whole lifetime to get to know this kid. Go to the hospital tonight but get back to work tomorrow." Henry hung up the phone.

Immediately, George left to meet Debbie at the hospital. There, he was asked to remain in the waiting room. The birth was not going well.

After hours of pacing the floor, George's worry was interrupted by a doctor and the new father was finally invited to meet his son.

In the maternity ward, George found his wife in poor shape and alone, the baby – a healthy boy – taken to the nursery to be cleaned up and bottle fed. Debbie had suffered significant internal trauma. She would need multiple surgeries in the weeks and months that followed. George was shocked at the sight of her as he sat in the chair at her bedside and held her hand. Debbie looked as though she'd been through war: her face was swollen and bruised from the violence of her ordeal and her eyes were bright red with the tears she had since cried.

"They won't let me go home, George," Debbie said, voice catching with grief. "They say I'll have to stay for a least a week, and that the baby will have to go home with my mom."

George gently kissed his wife. His heart broke to see her in such pain. Debbie was a woman who was meant to be a mother, and he knew what dreams she had of nursing her tiny baby, rocking the newborn in his new nursery, bathing and caring for their son herself.

"You need to listen to the doctors, Debs. The baby is fine. Your mom will take good care of him. All that matters is that you get better."

Debbie's mother had stepped out to give her daughter and son-in-law privacy, but soon returned with the baby in her arms. Debbie and George named their son Boyd, and George had handled grenades with far less nerves than he felt holding his infant for the first time. Boyd was healthy, George was told. Strong, but hard won. Yet, to George, the child seemed impossibly small. It was unreal that this tiny thing was all its own person, making small mews and squeaks, burrowing his eyebrows in concern while he slept. George was overcome with pride, love, and fascination in his son.

"You did so great, Debs," George whispered, nearly overcome with wonder. "Look at our boy. He's perfect. Debbie, you did so great."

As he lowered the child to Debbie, he watched her breathe in the scent of her perfect baby as she kissed him, and George realized his wife was the strongest person he had ever met.

He stayed until Debbie fell asleep, counting down each hour until he knew his next shift started, his resentment towards Harmony Helicopters growing with each passing minute. He thought of the money he'd secured for a company he had no real stakes in other than modest employment. He thought of every hour he'd worked for free to make something of the Slave Lake base, and of every machine he'd washed down to save Harmony Helicopters the cost of paying someone else to do it. And for what? To leave his injured wife and newborn son alone in a hospital room? All George was to them was a grunt. Before he left, George swore that prick Henry Stevens would never come between he and his family again.

He just had to figure out how to make that promise a reality.

FAR AWAY HELICOPTERS
1984

George had been working for Harmony Helicopters for fifteen months before he was presented with a new plan.

It happened one evening when Willy Light asked George to come for dinner at his house. The two had developed a casual friendship based on their professional interactions, their shared stage of life, and perhaps a trait they shared: unlike most men in the small town, neither drank alcohol. Willy had once had a drinking problem but had been sober for many years. He and his wife had also welcomed their own child just a few weeks before Boyd was born.

But Willy was crude. He was rough around the edges and uneducated, but was likable and friendly, and never, ever went anywhere by himself. It was as if Willy moved through the world fearful of solitude, as though he constantly needed the safety that came from companionship. He opened business after business, but had other people run them. Then again, Willy could afford to have others manage his businesses. Stepping into the mansion that was Willy's house, George couldn't argue that the man must have done something right, even if it was largely to marry into affluence.

"So," George said, as they dined on the expensive steak dinner Willy's wife had prepared, "I can't guess what business you wanted to talk about, but I can say I'm intrigued."

Willy lit a smoke as he pushed his plate away. "I don't want to feed you names anymore, George," he said. "I keep connecting you with customers, and those customers are making Harmony Helicopters rich, aren't they? They're not making me any more money. Not you, either. So why are we doing it?"

"What do you mean?"

"Well shit," Willy said, hulking in his seat to lean closer over the table. "It's not like I don't got the money. I don't want to pass on those clients, I want to keep them. Why don't I open my own helicopter business? Then you could bring those clients back, and you could work for me."

George was gob smacked. The wheels in his head were immediately turning. He knew the clients he'd secured for Harmony Helicopters didn't care about the logo on the side of the craft; they trusted George and hired him by name. He knew every account he'd secured would come with him if he left.

"What would this partnership look like?" George asked.

"Woah, slow down," Willy said. "Not a partnership. You'd work for me. You'd be in charge of the hanger and the flying. You'd bring over Harmony Helicopter's existing clients and I'll bring you more. You come help me buy a machine, and we'll get to work."

George didn't have to think about it long, but he did have a few conditions.

"I'm interested," George said, "But first, I have to be part of the company. I want to buy my way in."

Willy nodded. "Okay," he said, "We can make that happen. What else?"

"Two," George said, leaning in, "is that you are going to buy an A-Star, at minimum. Not a jet ranger – they're too light for the work we're going to do. An A-Star," George repeated.

"What's the difference?" Willy asked.

George laughed. "Think of it as the pick-up truck of helicopters."

"Well, we can do that. Is there a third condition?"

George nodded. "I need to discuss this with Debbie."

Willy Light grinned. "Of course."

By the time of George's dinner with Willy, Debbie had been released from hospital and the family had moved into the little apartment Harmony Helicopters had provided. Boyd was sleeping when George got home, so the new parents were able to give full attention to the opportunity at hand.

"We'd need to buy our way in," George explained. "I don't want to just be an employee anymore."

Debbie nodded. "Well, we have about forty-thousand in the bank," she offered.

"Yes," George said, "but that is meant for a house." The couple had scrimped and saved for their first real family home for years.

"But if this is successful," Debbie countered, "we could make back the investment and more."

She smiled in effort to show her sincerity, or perhaps to ease George's concern. "And I trust you, George. You've built Harmony Helicopter's Slave Lake base into the most successful location they have. If you think this thing with Willy is the thing to do, then we should do it."

Knowing he had the full support of his wife, George had some thinking to do. Ultimately, it was decided that the Kelhams would divide their savings, investing twenty-five thousand dollars into Willy Light's new business, leaving fifteen thousand to put down on their first home in Slave Lake. George would be considered a shareholder as well as an employee, making a return on the investment while also being paid an annual salary – modest, in the first few years, of course.

Once these decisions were made, things moved fast. They had to, so Harmony Helicopters and Henry Stevens wouldn't cotton on to what was developing under his nose. Together, they decided to call the new business Far Away Helicopters[9], and George took time off work to travel with Willy and Willy's accountant to England in autumn of 1984, where they bought Far Away Helicopter's first A-Star. Before they knew it, work was coming in booked months in advance. The Kelhams bought a little blue house in Slave Lake with what was left over after investing in Willy Light's new venture, and finally, it was George's last day under Harmony Helicopter's employ.

But Henry Stevens didn't know that yet.

The morning of his last day with Harmony Helicopters, George cleaned out his office. He left the keys to the A Star and the pick-up truck they'd lent him on the desk, and then made a phone call to Henry, telling him he was leaving. Then Debbie and Baby Boyd picked George up in the family car, and George

[9] Company name fictionalized.

delivered her and Boyd back to the house before heading to Willy's new set-up, beginning his first day at Far Away Helicopters.

While George was thrilled to be out from under Henry Stevens, he was no less busy flying for Far Away. They scored contracts with so many companies that George almost lost count. He often flew for Alberta Health Services, which meant being on call throughout the day and night to deliver nurses and emergency medical technicians to crisis scenes and transport patients from rural locations to city hospitals. Once, George flew a Bell 212 to pick up burn victims. The men's shack near Nipisi, Alberta, had filled with natural gas without their knowing and the very roof had blown off when something ignited inside. Another time, he was called out to a job site where a man unloading pipe was crushed after the massive amount of metal gave way. George knew from the sight of the man's injuries that his leg would be lost, and he thought of Tony's injury those many years ago. Later, he was sent with a registered nurse named Emily Elliott to Peerless Lake, assigned with the task of flying the corpse of a man who had overdosed after drinking photocopying fluid with two of his buddies. After that call, George flew a paramedic named Murray Parrot to a rural reservation where a young mother and child urgently needed medical help. The woman had barricaded her home's door with her body to keep her abusive husband from entering, but the man had blasted a shotgun round through the door, the weapon's pellets ripping through both the woman and the face of the newborn baby she held. The infant boy's name was Greg[10], George learned, and as he flew Greg to Slave Lake hospital, George sent a prayer up on the child's behalf to whoever might be listening. What a damned awful way for a child to start their life.

These events, as busy as they kept George and as emotionally charged as they could be, meant a great deal to George. George found fulfilment not only in building the business, but in helping people. He was finally 'that guy,' he realized; the man who showed up to an ugly situation and got people out of it, back to safety. Willy Light was more interested in his other businesses, which were already established and humming along, so complete responsibility for Far Away's success or failure fell to George, but

[10] Name fictionalized.

it didn't matter. George was used to working under heavy workloads at breakneck pace. If anything, owning ten percent of the company simply gave him more drive. Every contract he closed was not only paying his wage but was also an investment on future returns. He and Debbie had officially depleted their savings, having bought a home and into Far Away and now they had a child to provide for, to boot. George had never been so motivated to amass a rainy-day fund.

There was only one problem with the situation. George hadn't signed any papers and had little proof of his investment in Far Away. Whatever agreements had been made were agreed upon in verbal contract, which was a norm for the wealthy, meagerly educated, gruff men of northern Alberta, but the arrangement made George nervous. However, every time George thought he ought to bring up the issue, new work would come in that would all but run him off his feet, or family life demanded his attention. More and more, George was kept busy by Willy, who used the helicopter and George's services as though he was a personal chauffer. George's days off were spent delivering Willy to check up on the other projects he had going on throughout the north or dropping him into the middle of the bush to visit far-flung friends. George wasn't paid for this overtime work, but the outings blurred the line between pleasure and networking, and George often felt unable to refuse the jobs. Life moved like a river in fast current, and the task of insisting Willy draw up and sign a legal contract was one that fell to the bottom like a small stone, resurfacing every now and then upon disturbance. When this happened, George would remind Willy that paperwork was needed, and Willy would offhandedly make a promise.

"Sure. Sure, we'll do that."

But nothing ever happened.

At home, the Kelhams were preparing to welcome a new member of the family. Debbie was pregnant again, and the child would be born by scheduled c-section as natural childbirth was too dangerous following the trauma of Debbie's first birth. Because they were able to set a time for the baby's delivery, George was able to make sure his schedule was free, and the young father was thrilled to attend his child's birth. He sat in the operating room with Debbie, gowned up and holding her hand through latex gloves. A large sheet was strung up to prevent the parents from viewing the surgery taking place. They waited in silence for the surgeon on the other side of that sheet to perform his task, both excited and George nervous, Debbie numb from the epidural.

Finally, the doctor spoke.

"What is it you're hoping for, Mr. Kelham?"

George laughed through his mask. "As long as it's healthy, doctor!" he said. "But, if I had to choose, I'd love for our son to have a brother."

At once, a child appeared over the sheet, held high in the surgeon's bloody, gloved hands. The tiny baby cried with strong, clear lungs, gulping for air in the new, cold world it had found itself brought into.

"Then congratulations, Kelham family," the doctor said. "You have another son."

George laughed and laughed and kissed Debbie again and again. It was like the doctor had simply opened a present and pulled Baby Taylor out of a box, their boy a long-awaited prize.

Taylor's birth had been a total success. Debbie was comfortable and in good health, and able to nurse the little one – a facet of motherhood that she'd been denied prior, due to her pain and the strength of the drugs she'd been required to take after her first experience. George left the infant and new mother in good care of the hospital and drove back to Slave Lake to share the happy news of baby Taylor's arrival with Boyd, who was in care of one of Debbie's friends.

Debbie and Taylor spent a few more days in the Edmonton hospital while Debbie healed from her cesarean. When mother and child were ready to be discharged, George and Boyd flew to the city on Far Away's A Star and landed the helicopter right on the front lawn of the hospital. Then, all four members of the Kelham family flew home together, the first flight of young Taylor's life being the very same day he'd first felt the sun on his face.

~

Life didn't stop for a minute in those years. Between work and children, Debbie and George were run off their feet. With George pulling more than fifteen hours at the office or in the sky most days, Debbie handled almost everything at home. It was an area she excelled in, and she smoothly steered both children, casual employment, and all domestic responsibilities until their routines were bumped off course – a problem that happened more and more as Boyd, then a preschooler, developed painful eczema on his skin and began suffering uncontrollable asthma attacks.

Boyd illnesses had gotten so bad that Debbie was in possession of his doctor's home phone number – a measure that had turned out to be necessary more than once because of Boyd's dangerous condition and the fact that Debbie cared for both children so often without George's help.

One particularly scary night, Dr. Swartz had raced to the Kelham house as soon as he'd heard Debbie's frantic voice on the end of the telephone line. She'd all but desperately thrown her small child at the doctor when he arrived, Boyd's skin turning ashier and bluer with every passing moment. She wept as she watched the doctor cradle the boy, administering medicine in the backseat of his own vehicle as his wife, up front, pulled out of the driveway and raced to the nearest hospital. Debbie was left in terrified silence, Baby Taylor sleeping upstairs. But Debbie and George had good people in their lives, including Debbie's employer, Marilyn Dawson. As soon as Boyd left under his doctor's care, Debbie phoned Marilyn for help, and Marilyn insisted Debbie bring Baby Taylor over at once so that she might care for him while Debbie met her eldest at the hospital.

Debbie was at the doctor's office so often with the child that she soon became close friends with the couple who owned and ran the local pharmacy, Barton's Apple Drugs. Wendy and Dennis Barton entered the Kelham's lives like guardian angels. Fifteen years older than Debbie and George, they had extra time when the Kelhams had none, and they seemed to step into the role of surrogate grandparents for Taylor and Boyd.

It was, in fact, a relationship that could be said to have saved three quarters of the Kelham family. In 1988, Slave Lake suffered calamity. Rainwaters fell until the town flash-flooded, and no area was as badly affected as the southeastern quadrant that the Kelham's called home. That time, too, George was flying for Far Away, charged with picking up affected neighbors and delivering them to higher ground. Ironically, Debbie and the boys could have used the transportation themselves. While George flew, water filled the Kelham basement and soon seeped through the main floorboards of the upper level. The driveway was sunk beneath the swirling waters, and Debbie realized with terror she wouldn't be able to drive away from the flood. Luckily, there was soon a knock at the door – the town's fire chief had arrived on a boat to take them to safety.

Debbie and the children were delivered to a local elementary school where the town was housing flood victims on cots and mats. Debbie had only been able to grab a handful of diapers for Taylor and had no money for food or even a change of clothes for the boys. George heard his wife had been evacuated over the radio, but although he felt sick about it, he might as well have been a world away for how little he could help her while transporting others to safety. It was the Bartons who saved the day. They sought out Debbie and the kids at the shelter and brought them to their house, first stopping to stock up on infant supplies, food, and creature comforts. When at last George's work was done, the floodwaters had cleared. Their home was still structurally intact, but all else – photographs, furniture, appliances, and the large collection of hockey memorabilia George had been collecting since first discovering the game in Fox Creek – had been lost.

Holding his wife's hand as they surveyed the damage and juggled two small children to prevent them from playing in the muck of their nearly destroyed home, George felt terrible having not been there during the tragic event. Even then, there was little George could do to help other than hire contractors to clean up the mess. He had a full schedule delivering Far Away Helicopters' clients and shuttling Willy all around the north. In a very short time, George had developed Far Away Helicopters into a multi-million-dollar business, but he had yet to see any proof of his own success. All he had was a wage barely higher than what he made a Harmony Helicopters, and a workload that didn't allow him to attend to his family. More and more, George and Debbie questioned the legitimacy of the rights they had within the company. Were they truly shareholders, or had Willy swallowed all evidence of their investment? Was George a partner, or simply a manager? Right then, it was hard to feel as though he were anything more than an employee.

Resentment grew.

~

On the other hand, the thought of having been duped by Willy was so awful it made George sick to think about it. He learned too late that the man could be crooked in business, and he and Debbie had already sunk so much time and money into the company. They had also invested into the town itself. Debbie and the boys had made a home of Slave Lake, and George had built such success that it could have been argued Far Away had a monopoly on local helicopter aviation. To leave Willy Light's employment would likely mean having to leave town. The situation seemed impossible. Often, it felt like the best thing for George to do was simply to keep his head down and work.

One frigid Saturday morning in the middle of a winter cold snap, Willy wanted to go flying. George met him for breakfast at the Sawridge hotel and, afterwards, they drove downtown, towards Far Away's hanger. It was so cold outside that the street was clear of traffic, all residents of Slave Lake taking shelter in their warm homes.

All, except one.

In the distance, George and Willy saw a lone figure stumbling in the snow. As they drove closer, they realized the man had no coat, only a thin jacket, and he was wearing polyester pants and shined dress shoes which caused him to slip and fall in the crusted snowbanks.

"Jesus Christ," Willy said.

"Too cold for him to be out in that."

Willy drove towards the man. George thought perhaps the pedestrian was confused, near hypothermic, and possibly in need of help.

"It's Bert[11]!" Willy exclaimed as they pulled to a park beside the man.

Bert managed the Sawridge Hotel, where George and Willy had just had breakfast. A Kenyan of Pakistani descent, Bert was perhaps the only man in northern Alberta who had any skill in hospitality or hotel management. He not only knew how to meet client needs, but how to anticipate them. Bert had almost single-handedly made Slave Lake's only hotel a tourist's oasis in the middle of a rural oil town.

Willy pulled up alongside Bert. He rolled down his window.

"You crazy man!" Willy shouted to Bert. "What are you doing out in this cold?"

When Bert turned his head in George's direction, it was clear

---

[11] Name fictionalized.

the man had had a terrible night.

"Get in," George said to Bert. "Come on, we'll drive you home."

Bert lived just around the corner from George and Debbie, but where their home had not been entirely taken by the flood, Bert's had been completely totaled. Fortunately, insurance had covered much of the financial strain, but George could only imagine what hardships the event had brought on the young family.

Willy drove a single cab truck, so George slid over on the bench seat and Bert slowly slid into the vehicle, trembling with cold and crestfallen in spirit.

"I don't want to go home," he said. "I can't go home right now. My wife can't see me like this."

George looked at Willy and Willy raised an eyebrow.

"Where are you coming from?" Willy asked.

"The hotel," Bert said. "They fired me last night. They just called reception and canned me, no meeting or nothing. They want to give the job to someone else, I guess. I don't know how to tell the wife, so I just hung around the restaurant for the night. I didn't want to drive home after drinking, but I didn't realize it was so cold."

George was shocked. "They fired you?"

He knew Bert was paid modestly to begin with, and probably for only an eight-hour shift each day. However, Bert seemed to eat, drink, and sleep the hotel. He probably spent more time there then he did at his home. The hotel was likely making a fortune off Bert's labor, what on earth would convince them to let him go?

"I'll tell you what, Bert," Willy said, "We're not going to take you home yet. You stay with us for a while, and we'll take you for a helicopter ride. See what we can do to cheer you up."

The three men drove to the hanger and George took Willy and Bert up into the sky, the chopper warm because of the toasty dashboard heater despite the frigid temperatures outside. Willy wanted to visit a friend who was out on Peerless Lake, net fishing on the ice. They flew until Willy spotted a familiar truck parked by an ice fishing shack.

"Let's set down here, George," Willy said.

George landed the machine on the ice. He and Willy, better dressed for the weather, left Bert in the warm machine while they knocked on the door to visit whoever was inside.

"Come in!" a voice shouted.

The men entered the shack to find a man seated on a lawn chair, overseeing the equipment used for net fishing under the ice. Alfie was in construction, and often hired by Willy to subcontract. Alfie offered each man a beer, which George and Willy declined. George explained they couldn't stay long because of poor Bert waiting for them in the helicopter, and the man was interested in hearing the story. When George got to the part of Bert's dismissal, all three men were again rightly outraged by the tale.

"He was fired?" The man whistled. "Short sighted by the hotel. No one in town can work like Bert. Certainly there's no one around Slave Lake who knows how to run a hotel. What if he just takes up elsewhere? That alone could destroy the Sawridge's business."

"Well, there's nowhere for him to work, is there?" Willy Light laughed. "The Sawridge is the only hotel in town. Bert will have to sell his house, move elsewhere."

Alfie went silent for a moment. "Well, shit," he said. "It's not like you don't got the money, Willy. Why don't you build Bert a hotel?"

George couldn't believe what he was hearing. He wasn't sure if Alfie was joking, but either way, Willy had a thoughtful look in his eyes.

"Maybe I will, Alfie." Willy said. "Yeah, maybe I will."

Instantly, George knew the idea would come to fruition. Willy would find a few other funders and they'd build a large hotel which would become the Sawridge Hotel's only competition for years. Bert would manage it and, together, they'd take all kinds of work from the Sawridge. The hotel would open with sixty rooms, and work would be so successful they'd have to add another forty within a year.

The hotel should have paid Bert off, George thought. They should have made him sign a non-compete contract, and paid him to leave the industry or, if not the industry, to leave town. Instead, they'd sent the best and only hotel manager in town right through the doors of the competition – competition they hadn't even known they'd had.

George thought about it the whole way home. How shortsighted to not see the potential for one's own ruin in the skills of another man.

~

In 1992, life was a boiling pot.

Tensions were high throughout the country, as well as in George's life. In the east, Brian Mulroney's federal government faced bribery accusations, and Ontario was gripped in fear over what seemed to be a serial rapist and murderer on the loose – a man the public would later know as Paul Bernardo. Up north, in Yellowknife, violence was brewing in the form of a miner's strike. In little old Slave Lake, Alberta, the professional relationship between George and Willy had more than soured.

Their company was an inarguable success. Willy Light had bought more helicopters and more men were hired for George to manage, but George's wage had not increased, nor had any proof of his partnership been established. It was an issue George brought up more and more, and one that Willy consistently swatted away as though it were a pesky fly whenever mentioned. Perhaps feeling the increasing bad blood, Willy stopped asking George to fly him around so much. Then again, George didn't have time to hobby fly, even if Willy had wanted him too.

Far Away Helicopters had been hired by Peggy Witte, the large, hard-nosed mining executive who had bought Giant Mine, where the strikes were continuously ongoing due to Witte's refusal to allow the miners to unionize. Local helicopter companies hadn't wanted anything to do with flying scabs in, but Far Away took the job. Witte contracted them to fly in replacement workers from elsewhere in Canada to temporary camps in Yellowknife.

Immediately, it was obvious to George that this was a perilous position to be in. The strike was becoming increasingly ugly, and the picketers often crowded the helicopter pad to keep George from landing. Eventually, the RCMP got involved and George was given some sense of security as he brought in groceries and even delivered camp mattresses, removing the doors of his craft to fit the beds in horizontally. But the situation got worse and worse. The miners were strong-willed, hard men. Soon shots were fired, and people were injured. Eventually, a bomb was set off in Giant Mine, and nine underground miners were killed. It made headlines as one of Canada's worst mass murders.

After that, George was tasked with flying the RCMP Emergency Response Team. The members were made so large in their protective tactical gear that they seemed like guerillas, barely able to fit inside George's craft. On the next tour, George was charged with the responsibility of flying gold bars from the mine's head office to an Air Canada plane. The bars were so heavy George could only fly eight at a time, a single security guard sitting in the seat next to him. Finally, a month after beginning the contract, the miner's strike ended, and George returned home.

He came back exhausted and angry. The job had been a long one and so highly charged with bad feeling and litigation that it brought up the bad feelings he had around his dealing with Willy Light.

The next time Willy visited the office, George spoke up about it. This time there was no putting it off for another day.

"I built this company," George said. "Debbie and I put more than half of our savings into it, and I've personally flown and connected with every single customer we've had. You agreed to make me a ten percent shareholder, but I've never seen evidence of this. I want a raise, and to be made a partner – not an employee."

"Fine." Willy growled. "I'll have the lawyers draw something up. We can talk about this later."

George refused. "No. You're not blowing me off this time, Willy. I want to see proof of my shares, now."

But Willy Light was conceited. He sneered at George. "What are you talking about?" he asked. "You don't have any shares."

It felt like the room began to spin. George's shares should have been worth triple the money he'd originally put in – he'd worked so hard to make them so – and now Willy was saying they didn't even exist. George felt as though he might throw up. He felt as though he might explode.

"Listen," Willy said. "We can talk through this. You're overworked. Go home, take a couple days off. Then we'll talk again."

George went home feeling sick to his stomach. He consulted with Debbie. What would their next move be? Neither knew what to do.

The next day, George drove to the hotel and conference center Willy had built to have Bert run. He found Willy drinking coffee in a window seat at the restaurant.

"We need to talk about my shares, Willy," George demanded.

"You contributed seed money, George," Willy lied. "There are no shares. But if you want your money back, fine. Drop in the bucket, for me. How much was it again? Fifteen thousand?"

"It was twenty-five," George said in horror as Willy sat there, sipping his brew. "And you and I both know it wasn't seed money – I bought my way in. Those shares would be worth more than fifty grand by now!"

Willy leaned back and narrowed his eyes. "I don't know that at all, George. I'll have my assistant deliver the cheque." He shook his head slowly, and grinned.

"Twenty-five K," Willy whistled. "What will you do with the money, George?"

The sound of blood was rushing through George's ears. He thought he might black out with fury.

"I'm worth more than that, Willy," George said, struggling to keep his composure.

Willy laughed, again. "Alright. You're right! You helped me grow this company from nothing to a multimillion-dollar enterprise. Tell you what I'll do. Just because I'm generous, I'll tag on another ten grand. What does it matter to me?"

George left the hotel. He stood in the parking lot and didn't know what to do.

He went back to his office at Far Away Helicopters because it was his job, but later that day, the phone rang. It was one of Willy's cronies.

"Where are you?" George asked. He knew the man had been in town that very day, but the call ID read that he was in Edmonton.

"I'm in Edmonton. You're fired, George," Willy's sniveling yes-man said, over the phone.

George laughed. "You raced back to Edmonton just so you wouldn't have to say this to my face, didn't you? And Light didn't even have the balls to do it himself."

They were cowards, both.

George went home to Debbie. She was surprised to see him in the middle of the day and immediately knew something was wrong. The children were at school and the house was quiet.

"Willy fired me," George said.

Debbie didn't believe him. She shook her head and laughed at the joke. "Whatever," she said, before realizing he was serious. "George…. no."

But George simply swallowed hard, his eyes glistening with the tears of anger he withheld.

"Oh my god," Debbie whispered. She crossed the room and wrapped her arms around him.

"We'll be okay," she said, breaking the embrace. "Listen, we've always been careful, right? We don't waste money. We still have the three-legged couch, for heaven's sake!" Debbie laughed, trying to lighten the mood. "Being careful has made us able to pay off the house. We're not in an awful position right now, and you've always taken care of us, George. And I have a job. And you'll find another."

George just shook his head. His throat was so tight with grief he couldn't make words.

That evening, Dennis and Wendy Barton came over to console the Kelhams. George appreciated their kindness and was comforted by the love of his wife, but he felt like a failure. Debbie had already been taking care of the kids and the house and working part-time bookkeeping – mentoring under Marilyn – and now, because of him, she was going to have to go fulltime while he was a nobody. A good for nothing.

She'd be better off without me, George thought. Debbie and the boys would have extended family, and Dennis and Wendy, and many more friends, too. It would be better if George was dead, he thought, not for the last time.

Refusing to take from what little savings he and Debbie had after paying off the mortgage, George found himself an entry level job driving a gravel truck the very next day. At that time, anyone with blood circulating could find a job in a minute, but the pay was far less than the family had become accustomed to. He made twelve dollars an hour at Emil's Right of Way and hated every minute. It was humiliating. He'd gone from chief pilot of a successful helicopter company to a nobody – a mere machine himself, completely replaceable. In the weeks that followed George's firing, suicide was a thought that became more and more prevalent, entering his mind even as he tried not to dwell on such dark ideas. In town, people he'd worked with now snubbed him, perhaps to avoid landing in Willy's bad books and losing work, themselves. The cruel self-talk in George's head ran as though his mind was a compact disk left on repeat:

I've let the family down. I've failed. I'm no good. No good for anything. They're better off without me.

Debbie was concerned. She seemed to intuit what beliefs he was battling. Daily, she reminded him of the many different ways in which he was important to her. She told him to keep perspective, that he was more than a job, more than a pay cheque, more than even just a person - that he was a father and a husband and was loved completely. But while her words touched him, George couldn't accept them. At the end of each thought, he was reminded that he was *her* husband; he was this wonderful, resilient, grounded, kind woman's man, and he was letting down the best person he knew. He was supposed to be providing for her, and for their children. They were everything.

He was nothing.

With Debbie at work more often, it fell to George to spend more time with the kids, but he wasn't really *with* them. He struggled to stay in the present, in an attentive state of mind. The boys were school aged then and enjoyed motor cross racing on dirt bikes just as their old man had when he was a youngster. Every day, George would take them to a racetrack and stand on a small hill, watching them ride laps below. No one would have thought that proud father was planning his suicide as he stood there, watching his children, but he was.

Then, at his very worst, George's dark thoughts were interrupted by a man he barely knew.

The man's name was Dwayne Taylor, and he and George were casual friends. That day, Dwayne climbed the hill George was standing on, and smacked George's back so hard in greeting that it caused George's body to sway in apathy, as though the act of standing against the force of Dwayne's hand took too much effort. It was maybe in that gesture that Dwayne saw some insight into George's mind because at that moment, the man suddenly became gentle and kind.

"It's going to be okay, George," Dwayne said. George looked up and saw Dwayne was sincere. "You're going to be alright. You'll get through this."

The kindness of one man, nearly a stranger, was powerful. It was the sign George needed, and part of him was reminded of that sense he had felt so many years ago, of still being protected by the love and prayers of his mother. Had Louise Susannah sent Dwayne to say what she could not? For the first time in weeks, George remembered who he was. Yes, he would get through this. He would make it alright.

For a season, George found flying work. He and Leroy Dean piloted helicopters throughout Greenland, over the arctic country's flat ice plate and barren, rocky coast. The work was invigorating but short-lived, and George was contracted to work only one summer before returning to Canada to drive truck.

George worked as hard driving the gravel truck as he had piloting helicopters. He had to – it was a promise he made to himself. Willy Light might have clipped his wings, but George wouldn't give him the power to change his character. He decided if he were to be a gravel truck driver, he'd be the best truck driver he could be. He started work early, he left work late. He gave everything he could to his family in off hours, though there was little money to go around. He did this work for what felt like a long while then, in spring 1994, George received a phone call from an old flying friend.

The United Nations was actively recruiting helicopter pilots to serve in Somalia.

George was going back to Africa.

SOMALIA
February 1994

There was only one place George wanted to be less than in Africa, and that was the ground. If flying meant George had to return to hell, it was something he was willing to do, especially if it meant he could put food on his family's table.

He had asked Debbie for her blessing first, of course. She, being brave and generous, never even considered asking him not to go. The idea was that George would repeatedly fly in Africa for a tour of six weeks each time, and then be returned home for a month of rest. The money, while nothing to write home about, was an improvement on what he was making driving trucks and would not be taxed due to the Canadian government's commitment to aid the United Nation's missions by honoring the sacrifices made by UN workers.

When it was finally time for George to begin his first mission, he said goodbye to the children in the morning before they went to school, and then Debbie drove George to Edmonton, where there were Air Canada plane tickets waiting for him at the ticket counter. At the terminal, George kissed his wife goodbye and told her how much he loved her, and promised to be safe, and it wasn't until after he boarded the flight to London that he realized Debbie had never once spoken her concerns or complained how she would manage parenting and working while he was gone. She was a warrior, George thought. A woman like no other.

When the plane landed in London, George was transferred to

a flight headed to Nairobi. Once in Kenya, he flew to Mogadishu, Somalia. The Russians had won the contract to move UN staff into Somalia, George found himself boarding a clunky Tupolev airplane, large enough to fit 180 passengers.

George found an empty seat in the dark craft filled with people and sat down, placing his backpack at his feet. When his feet finally touch the Somalian runway, a man stood outside waiting for him, his hand outstretched.

"How did you enjoy the sewer-tube?" he joked in a thick French-Canadian accent, nodding towards the Russian plane and making a play on its name. "Long, skinny, and smells like shit."

George laughed and extended his hand. "George Kelham," he said. "Canadian?"

"Oui. Quebecois." The man tried to place George's accent as he accepted the handshake. "South African?"

"Rhodesian, originally," George said, shaking his head. "So, Zimbabwean, I guess. Canadian for the last ten or so years, anyway. I'm a pilot. Here to fly."

"Ahh," the man said. "Wonderful, welcome. My name is Richard Barrett. I am glad to have met you. We will work together."

It turned out that Richard was also a helicopter pilot. George soon learned there were three of them, George, Richard, and another Canadian who went by Bryan Arsenault. Their supervisor was named Brian Clegg, and he oversaw the Canadian operation. They were working as subcontractors under the UNOSOM umbrella.

Richard was the first to show George the ropes when the Russian plane brought them to rest on the UNOSOM airfield. There was a sense of safety in the camp then; although a chaotic war was raging at all parts of the camp's border, the US Marines had a very strong presence and had occupied one side of the airfield. UNOSOM's operation occupied the other. Medics, most hailing from New Zealand, were granted special permission to carry weapons even while in camp, and their presence added a further layer of protection. American CIA quarters were also setup nearby, and that created somewhat of a triangle of safety. A short drive away, inside the city of Mogadishu, the Pakistani military had set camp in an abandoned soccer stadium, but George was never taken to see that place. He just knew of it by the Pakistani soldiers who would travel to the airfield to be transported by George, Richard, or Arsenault.

The Canadian pilots were set up in temporary trailers in a small camp on the east side of the airport runway. Each pilot had their own room and shared a common recreation area. As the operation got larger and required more pilots, George would work with one or two Aussies, Kiwis, and even an American, but ninety percent of the helicopter pilots were Canadian, as Canadian Helicopters had won the contract. George met another Albertan named Gordie Bean, and the shared home and culture of the men were a small cure for homesickness. Unless in helicopters on assigned missions, the Canadians were never to leave their camp. All they had were one another.

This set up was not easy at the start. In George's first days in camp, he realized he was getting a lot of side eye from other men as they passed from their trailers to the runway.

"What's their problem?" George asked Richard, gesturing towards the unfriendly strangers.

"Those are the Russians," Richard explained. "They had this contract before Canada won it. We are taking their jobs, one at a time. Their pilots are transitioning out."

"Ah," George said. "Well, I'd be pissed at me, too!"

It soon became apparent how bitter the Russian contingent were. Every day, George would hear words spat at him in Russian, see angry men kick over equipment, or get the sense daggers were being shot into his back by the eyes of pilots glaring at him as they smoked cigarettes. But it was also obvious why Canada had won the contract. While Canada's machines were smaller than the Russian's, they were certified and the Russian's old clunkers were not, plus the Russian pilots themselves lived in a state of mind every bit as outdated as their machines. Their bitterness came to a head one day when an American Blackhawk set down on the tarmac. As the shining, lightweight, nimble craft lowered to the ground, the Russians stumped out the cigarettes they held between their index fingers and thumbs and spat in the Blackhawk's direction. When the pilot exited and removed their helmet to expose a head of long hair and the delicate face of a young woman, the Russians howled in rage.

It was best they were leaving.

George was assigned a Bell 212, a helicopter originally meant to carry fourteen passengers, but that had had its seating removed to accommodate auxiliary fuel tanks. His mission was to fly medical evacuations and casualties to the Pakistan hospital, as well as to transport negotiators to the leaders of fighting factions for peace talks. When flying into dangerous airspace to meet these warlords, George was accompanied by a second helicopter to create a semblance of protection for the negotiator. Of course, it was all for show. Had a Surface-to-Air Missile (or SAM) been shot into the sky, two choppers could go down as easily as one.

The idea was that UNISOM was going to bring peace among the fighting factions, but George already knew better. Men who love war have no want for peace, and no regard for human life. George got his hands on excess bullet proof vests and lined the floor of his helicopter with them, should he be shot at from below. He kept his eyes open, always, and tried not to let Debbie worry about just how dangerous things really were. George, Richard, Gordie, Arsenault, and the other pilots were working around the clock. Rapidly, more Canadians were brought in to help lighten their load, but just as soon as the new pilots landed, they would pack up and return home. Most cited that the mission was just too dangerous. They were right, it was.

But in Somalia, George was a pilot, and not a truck driver.

The nation was in the grip of a civil war. The Battle of Mogadishu had taken place only months before George had joined the peace efforts, and the bodies of American soldiers had been jubilantly dragged through the city's ruined streets by the supporters of Mohamed Farrah Aidid. George had watched news footage in shock and disgust with Debbie, at home. Now he found himself thrown right into the middle of a world that had seemed so far away it might as well be on another planet compared to peaceful Canada. George was even charged with transporting Aidid himself to peace talks, flying with a person one might have considered the most dangerous and atrocious man in the world.

George had seen how little life meant to African warlords, especially female life. When not involved in the transportation of those involved in peace talks, George would assist the UN's humanitarian efforts. Sometimes this meant rescuing young girls from female circumcision rituals – some of which that had an over fifty percent chance of costing the children their lives due to the recurrent infections they'd experience after the mutilations, never mind the danger of childbirth, or even the risk that would be associated with sexual interactions later in life.

In a different mission, George transported the barely

breathing body of a woman who had been burned in a car wreck, the long skirts of her hijab withering beneath the flames and melting to her skin. In yet another trip, George flew to retrieve and transport a newborn baby boy. Thankfully, no evil had been done to the child who so reminded George of his own sons as infants. The baby simply suffered urgent health concerns. The child had failed to develop an anus in utero, and thus needed to be rushed for sophisticated surgery at the Pakistan temporary hospital in Mogadishu.

In this particular case, George left the helicopter to smile at the beautiful, wide-eyed child as a translator explained to the parents where George would take them and their infant. It was difficult to believe this child had anything wrong with him at all; he looked brand new and perfect in the same way all newborn children do. As George smiled at the baby, he was relieved that the assignment involved something as positive as helping a family. At last minute, as he sat in the driver's seat, George realized he could hear the child's mother screaming. He looked behind him to see the baby's father pushing the woman away as the father boarded the helicopter with the baby in his arms. This caused the new mother to fall to the ground just hours after she had given birth, her arms outstretched to the infant as her husband prepared to travel with their son and leave her behind. There was room to take both parents, and the woman was welcome to come, George wanted to say, but there was no shared language between them. It wouldn't have mattered. George realized the man cared not at all for his own wife, but instead blamed her for the problems endangering his son – problems far, far beyond her control. All relief was stripped from George as he brought his craft to lift above the dry, yellow earth, the sobbing woman beneath them, curled into the ground in blankets of black skirts. Nothing in Somalia was ever fair.

Somalia was as much a hell as patrol of the Rhodesian flatlands had been, but, just as he had there, George kept his head down and his eyes open. He did his work until it was time to return to Debbie and the boys for his four weeks off. Once back in Canada, he would act like a father again, imagining he was just returning home from a safe oil camp, like most other men in Northern Alberta. George would play with his kids and watch hockey with his wife. He'd marvel at how she kept the family going without him, recognizing the routines she and the kids had forged in his absence. He knew it hadn't been easy for Debbie, but he also knew his return didn't provide much relief. Being away so often and for so long, George lost track of how the family moved. He would sit in someone else's unofficially assigned seat or make plans only to learn the boys already had obligations to fulfil. This bothered George, and he always felt a deep need to try make up for the time he had missed. While Debbie diligently ensured the children made every dental and eye exam, and never missed a day of school, George would sometimes break them out of their elementary classes just to spend time with them.

One day, when Debbie took George's return home as an opportunity to spend a couple nights away with a few friends, George bought concert tickets and had the principal's office call Taylor out of class. The little boy had entered the school office pale-faced and meek, looking terrified at the prospect of landing himself in some unanticipated trouble. George smiled at his young son.

"I've got three tickets here for The Rolling Stones," he told his boy. "Your brother doesn't want to come because he's got hockey practice tonight, so Leroy's going to stay with him. Would you like to see your first concert with me?"

Taylor's eyes were as wide as saucers. He leaned in to whisper to his father. "Dad," Taylor said, looking side to side, "I don't think Mom will like this."

George laughed. "Well," he countered, bending down to his child, "Mom's not here right now."

Determined not to forfeit his place in the community, George threw himself into volunteerism on his days off. He and Debbie opened their home as a billet for young hockey players, and George coordinated the local team's annual fundraiser pancake breakfast. On tours off, George drove *The Wolves'* bus to and from out-of-town games.

To acknowledge his volunteerism, the team's coach scored George and his family tickets to watch the Edmonton Oiler's morning skate before they played the L.A. Kings. It was a banner occasion for the Kelham family. George had been collecting Oiler's memorabilia for over a decade at that point, and while he'd lost part of his collection in the flood, he thought this might be a great opportunity to grab an autograph or two, if he was lucky, and the timing was perfect as George would be leaving Canada from Edmonton to Mogadishu the next day. He and Debbie loaded Boyd, Taylor, and an unsigned, never-released-in-Canada poster of Wayne Gretzsky and Gordie Howe that Deb's brother Mike had bought for them in America, and the Kelhams drove to Edmonton early in the morning, three hours before the players took to the ice.

Together, they watched the Oilers skate. On the way to the dressing room, Taylor, who was about eight years old at the time, frantically waved down a player, and when he turned to greet the child, George was floored to see it was The Great One himself! George passed his son the poster and Gretzsky signed it before continuing to the showers.

On his next days off, George received an even better gift. Between the duties of work, the busyness of Christmas, and the constant work of raising two boys on her own, Debbie had rented out the Slave Lake Native Friendship Centre and planned a celebration for over eighty guests when George had all but forgotten his own fortieth birthday. When a cake with lit candles was presented to him and family and friends gathered around the table, George's sons sitting on either side, he felt like a king. He had come from a family who had never shown such free-flowing love and acceptance. He had survived the darkest moments of his life – not only war, but the war of one's own hopelessness. Somehow, he was still in warring Africa, but he'd find his way through that, too.

If Somalia had been dangerous before the United States Marines packed up and left the country, it was certainly more dangerous after Bill Clinton pulled his remaining troops. George and Richard watched in dismay the day the Americans loaded their camp gear, jeeps, and even their helicopters in to massive C5 galaxy planes. Newer Canadian pilots, who had stayed until that point, put in their notices, and stated their immediate intention to resign from service. Without the marines to protect the land in front of the airstrip, warring factions moved in closer. One day, George and Richard were standing near their machines when George heard a loud crack over his head, and the rush of wind on his skin. Without even realizing what they were doing, he and Richard scattered like scared rabbits, crouching and running at the same time. Someone had shot at them, they realized. Who? They couldn't see. It didn't matter. The land was flat and barren. A rebel could be a mile away, and still there would be risk.

Morale was as low as risk was high. The international

community was beginning to withdraw their support from the United Nation's efforts. George felt he was in a holding pattern. The mission certainly couldn't continue forever, and it certainly had to end soon, and the last pilots there… well they would be the ones in real trouble. He didn't know if he ought to end his contract or wait to be sent elsewhere. He began to feel the familiar unsettling feeling he had during the Bush War.

An unwinnable war.

John Leonard's voice rang in George's mind like the words were spoken only moments earlier.

Finally, Clegg received word from head office. Two Canadian helicopters would be sent elsewhere to support the UN's efforts. George and Richard were selected. They would leave Somalia, for now.

Help was needed in Rwanda.

RWANDA
April, 1994

In war, small blessings are everything. For George, a small and important blessing was that his friend Richard Barrett would be stationed with him in Rwanda.

The pilots joined with an Aircraft Maintenance Engineer named Matt[12]. Together, they flew their helicopters in tandem to Nairobi, Kenya, and then to Entebbe, Uganda. After fueling up, they flew over Lake Victoria and into Kigali, Rwanda, where they met General Romeo Dallaire, the Canadian in charge of the United Nation's UNIMAR program.

At first glance, Rwanda seemed like a paradise compared to what George and Richard had left in Somalia. The people were friendly, and the men were put up in a hotel instead of a worker's camp. The pilots were given a vehicle for travel, and George felt almost as though he were on vacation. On their first night, the men unpacked their suitcases in their separate rooms and then met in the lobby of the hotel for dinner. George ended the night with a call to Debbie before falling into a deep, restful sleep.

The next day, George flew General Dallaire to a Hutu extremist camp for peace talks. Tensions were high, and George knew from his briefing that extreme prejudices had taken place even within Rwanda's general population, spurred on by new media such as radio. But while George was ever vigilant during

[12] Name fictionalized.

the General's transport, he didn't feel any danger among the citizens of Rwanda. In the hotel, the clerks were friendly. Those who worked in the building, those who delivered goods, and the other hotel guests, all seemed perfectly adjusted – certainly not consumed by the hate Hutu extremists were promoting within the populous.

The next evening, Richard scouted out a place to dine and the men enjoyed flavorful plates of food before returning to the hotel. Tasting the excellent, fresh food he'd been prepared, George realized he'd forgotten about some of Africa's most wonderful qualities. Rwanda had an undeniable beauty, and, in a small way, George felt as though the trip were a return to his childhood home. After the meal, the men said goodnight to one another in the hotel corridor, and George retired to his room. He was scheduled to provide transport for peace talks the next day and wanted to get to bed early to arrive well rested.

George was asleep when his wife called him. In Rwanda, it was early in the morning, but it was late in the day in Canada. George moved to the balcony of his hotel room to enjoy the cool air as he spoke to Debbie over the phone. Debbie was chatty and seemed well, and in the background, he could hear Boyd and Taylor quietly playing.

Suddenly, there was a loud sound, like the crack of a gunshot magnified many times over.

"What was that?" Debbie asked. She'd heard it too, through the phone line, all the way across the world.

"Ah… the natives are getting restless, I think," George joked. His eyes scanned the horizon and caught sight of a plume of black smoke that started in the sky and trailed downward. It was an aircraft, George realized, a plane or helicopter had exploded or been shot out of the sky.

"Debs, I gotta go. I love you, okay?"

"Okay, George. Love you too."

George hung up the phone to almost immediately hear a knock on the door. Richard was waiting in the threshold.

"What was that?"

"The Rwandan president's plane," Richard said, a dark mood passing over his face. A chill passed over George.

"Whatever you think hell is, we're about to see much worse here."

George invited Richard into the room. They had no radio and

no TV in which to find news, but they did have UN walkie talkies. From the walkies, they had it confirmed that, indeed, the Rwandan president's plane had been shot down. He and all crew members were dead. The Rwandan government had been overthrown.

George and Richard looked at one another. The president's plane had left from the same airport where their helicopters were parked, only about ten kilometers away from the hotel. George's hotel phone rang. It was their supervisor at the United Nations. George placed the call on speaker, so Richard could here, too.

"You are not to leave the hotel," their manager reported. "From this moment until you hear otherwise, stay in the hotel and, most ideally, do not leave your rooms. Belgium forces have been dispatched to provide security to the hotel. Can you relay this message to your coworkers?"

George confirmed that Richard was with him now, and that they would consult with the third co-worker on assignment with them.

George hung up the phone and sighed.

"It'll be over by tomorrow afternoon," George guessed. "We've just got to wait out the confusion."

Richard nodded. "Yes. Good thing we ate last night!"

The men bitterly laughed, and then headed down the hotel hall to report the situation to Matt.

~

The two pilots and their Aircraft Maintenance Engineer each slept in their own rooms that night. They were somewhat uneasy because of the events of the evening but felt sure of the security that guarded the hotel and certain that law enforcement would soon have the situation under control, possibly by the end of the next day. When the men woke, they convened in George's room and headed to the lobby of the building together, in search of information from the concierge.

The Canadian pilots weren't the only ones who had that idea. Most of the hotel's guests were there, and the hotel manager was encouraging everyone to stay in their rooms for the day. Breakfast would be provided, as well as clean towels. Outside the lobby windows, George could see Belgium soldiers standing at the door. He could leave if he wanted to, he realized, but there was nowhere to go. The airports were under control of the rebels. With great discomfort, George realized he was not a hostage of the hotel, but of the country itself.

As the men ate their food, George watched the other hotel guests, some, even, with small children. Afterwards, each person went back to their rooms, awaiting news, feeling more and more nervous that perhaps things weren't getting sorted as quickly as they thought they would.

George called Debbie, and she told him all she knew from international news reports. No representative from the UN had spoken with her yet.

"If I get out of this," George tried to lighten the mood, "I'm writing a book."

Debbie didn't find the joke very funny. "There is no 'if,' George. You're coming home."

After their conversation, George sat on the bed in silence. Richard and Matt sat on chairs in the same room. Then suddenly, like how the tension shifts right before an electrical storm, something in the air seemed to change. Something had let go. Something was very wrong.

The walkie talkie Richard had with him began to crackle. The hotel room filled with the sounds of frantic voices. The words were in French, and Richard translated the conversation for George and Matt, the color draining from Richard's face. Before Richard began his translation, George knew the speakers were Belgium soldiers. They were begging for their lives.

"We just want your weapons," Richard translated.

"It's the Hutus," Matt whispered.

The begging and shouting continued.

"Give us your weapons," Richard said, translating what the loudest, fiercest voice screamed. 'We will negotiate your safety." Richard looked up at George. "The other men are begging. They are saying, 'Please don't kill us. Please don't kill us."

The soldiers on the walkie began weeping, quietly at first, and then there was screaming. After fourteen gunshots, the line was empty but for the cracks of static.

George, Matt, and Richard sat in stunned silence.

"They've been killed," Matt whispered.

George's heart pounded in his chest. He turned and looked at the window, covered by curtains for any extra safety he could muster. What the hell was going on out there? Carefully, George used two fingers to part the curtain from the wall, so he could glance outside while keeping himself from view. Underneath his balcony, a heavily armed man – a terrorist – ran through the hotel's flower bed, trampling flowers as he stalked his prey.

It wasn't long before the day turned into night, and then the night became a chorus of screams, and gunshots, and chaos. Matt, George, and Richard stayed in George's room, not leaving even as the power shut off. Despite the screaming, they avoided looking out the window, lest someone see their figures move behind the curtain.

In the morning, George noticed there was no more water pressure in the bathroom. The men went down to the hotel lobby and met terror on the face of every hotel guest. The manager explained that war had broken out, and that there were mass killings in the streets. Neighbors had turned on neighbors. Women and children were being slaughtered by machetes alongside their unarmed men. The UN was negotiating the release of the hotel guests, but the parties had come to no agreement. Food would be rationed. Phone lines had been cut. There was no more water.

Richard's walkie talkie was running low on batteries. The men stayed in the hotel room, filling their bathtubs and sinks with whatever water was left in the line, rationing food, playing cards and staring at the walls, trying not to listen to the screaming outside, and then listening for screaming when the silence took over. Over the next few days, they left their room only for morning news in the hotel lobby.

Finally, after six nights, they came to the lobby to find a platoon of UN soldiers standing within its walls. The hotel guests and skeleton staff – the manager and a single clerk who had stayed behind – huddled in the center of the hotel's dark lobby, the space feeling almost junglelike to George by the camouflaged men guarding its periphery and the filtered light that found its way in through the boarded windows. General Dallaire himself had come to greet the hotel guests. He explained the situation; the country had descended into civil war, with civilians largely carrying out the carnage. The United Nations had negotiated the release of the Canadian pilots and other travelers staying in the hotel. The plan was that they would travel by convoy to the airport, which was under control of the Hutus, but guards would allow the hostages to leave the country.

The hotel guests were encouraged to quickly place themselves in a vehicle. George, Matt, and Richard were to take the pick-up truck assigned to them on their first day. Richard drove and George sat in the passenger seat, placing his small suitcase between his body and the door with the hope it might stop any bullet shot at him. He was scared shitless.

The world outside the hotel was bright but eerily silent considering it was the middle of the morning. George hadn't been outside for six days and it felt as though he'd stepped into an alternate dimension when compared to the sunny, hospitable city he'd known the evening the president's plane was shot down.

But for the silence, it wasn't until the convoy pulled onto the main road that George saw signs of the massacre. There were bloodied bodies everywhere. Murder victims lay on either side of the road, some unceremoniously pulled from the asphalt to the ditch, pools of blood documenting the move. Limbs had been hacked off, and hands and arms littered the ground like discarded garbage. George could hear gun fire all around them. In between the buildings they passed, women fled, tugging the hands of their crying children behind them. Some had given up on running. They simply knelt by the bodies of their loved ones, waiting for death to come for them, too. Waiting for their killer to arrive.

The convoy drove slowly. It was so hot that George sat with his elbow bent so it hung out of the open window until Richard told him to put his arm in as they passed machete-armed men.

"They'll cut your arm off just because they can," Richard said through gritted teeth. "Don't give them the opportunity."

When the vehicles finally arrived at the gates of the airport, George saw that the Hutus had set up a checkpoint. They were inspecting vehicles for Tutsis or Tutsi sympathizers. If a Tutsi – an innocent man, woman, or child – was found smuggled in the trunk of a car, or hidden under luggage, all in the convoy would be executed in the street.

The guards had blood on their clothes. They shouted at the drivers of the convoy and pointed guns into the cab of the vehicles as they pushed its occupants aside, throwing bags out of the way. They would kill us on the spot, George realized as he sat with his hands in the air, facing forward, hardly breathing.

"Bonjour," Matt stupidly spoke up from the backseat, as though thinking it were possible to befriend the enormous, terrifying Hutu soldier who readied himself to shoot at any moment.

Shut the fuck up, Matt, George thought, wishing he could reach back and strangle the man. Matt didn't even know how to speak French. Had the guard answered his greeting, Matt wouldn't have known what to say next.

Finally, the man in charge of the check point was satisfied. He nodded and the other men stepped aside and withdrew their guns, letting the UN convoy through. One man stared into George's eyes as they pulled away, but George couldn't imagine what the soldier could be thinking. Was he pleased his country had exploded into such chaos, and he part of the destruction? Was he relieved that his team had come out on top? Or did he know Tutsi people? Did he love a Tutsi person? Was that guard sickened by the outcome, and yet forced to play along to keep his own life? Or had the man simply given in to evil, as so many do?

There was no way to know.

Inside the protected airport, George and Richard finally exhaled. They were both shaken but it was not yet time to mourn or celebrate. First, the men needed food, water, and to bathe, and to rest. That would happen immediately. That very night, George and Richard were assigned to temporary living quarters, built in a modified Sea-Can shipping container. Next was the task of leaving Rwandan airspace, a more difficult plan to hatch.

The rest of the hotel's guests had flown into the country by plane, and so were being flown out by such as well. Matt went out on the first plane with the women and children, abandoning the helicopters he was meant to maintain. But George and Richard still had their machines in the airport. They argued that they couldn't leave such expensive equipment.

"We can't guarantee your safety," an official from the United Nations explained to George and Richard when the pilots insisted on flying their crafts home. "If you fly yourself out of here, there's every reason to believe you will be shot down by rebels."

George shrugged. That could happen to an airplane just as easily as it could happen to a helicopter. The only real problem was that the Canadian pilots had no fuel.

George and Richard set to work searching for fuel that evening, but it seemed their luck had run out. The men stayed in the temporary shelter for three or four more days as they continued to search for fuel. They should have been dog-tired after the ordeal, but they were still in survival mode, energized by their sole aim of finding safety. Eventually, Belgium military pilots were willing to share some of their fuel with George and Richard, but just enough to get them to Tanzania, where they could reach temporary safety and fuel up.

The Canadian pilots filled their tanks just full enough in the dead of night. They split what money Richard had on his person – about ten thousand American dollars – in case they got separated on the way to the fuel truck in Tanzania, and then, finally, they slept.

When first light arrived the next morning, George and Richard flew their choppers through the threat of bullets to safety, like bats out of hell. After they crossed Lake Victoria and landed in Tanzania, George exited his machine and lowered himself to his knees, his eyes closed, face lifted to the warmth of the sun. He was alive.

Palms faced down, George bent low, and kissed the earth.

~

Richard and George were debriefed in Nairobi. After six days since their last phone call – the longest George had gone since meeting his wife – George was finally able to call Debbie, but the phone just rang and rang. Knowing she'd be terrified for his safety, he called Wendy Barton at the pharmacy.

"George!" His wife's friend started to cry on the other side of the world. George cringed. There was no time for tears. The phone connection was already bad, and liable to fail at any moment.

"Wendy, I've only got a few minutes. Please tell Debbie I'm safe in Nairobi. I'm coming home."

"Dennis! It's George!" Wendy shouted. "George, was it terrible? It's been all over the news. Did you see what happened?" Every second word she spoke was crackled with phone line static.

"Wendy, please just tell Debbie I'm safe. Tell Debbie I'm in Nairobi."

"Where are you?"

"Nairobi. Safe. I got out of Rwanda." The line went silent. George had no idea how much Wendy had heard, and yet he needed to rest.

He joined Richard at the outdoor restaurant in the lobby of the hotel the UN was putting them up in for the next few nights.

The company they were contracted under were happy for their safety, but happier still that Richard and George had rescued the helicopters. Later, each pilot would receive letters thanking them for their courage in securing the aircraft, but there was no extra danger pay, and no extra time provided to them to recover from the ordeal.

The men were flown back to Canada for their tours off, and George returned home to hug his wife and children, trying his best to act like a normal dad.

GUATEMALA
1994

Facing increased international criticism for their role in the Somalian war, the United Nation's programs were being scaled back, phased out, and the Canadian helicopter company only had a few months left in the contract. For George's next mission, he and Richard were sent to Mozambique. The war-torn country was about to hold their first general election, and the UN had agreed to send ten helicopters across the many villages throughout the country.

The helicopters transported do-gooders from Europe, people who had agreed to oversee the election and ensure everything was legal and fair. The United Nations hoped this act would ensure the new era Mozambique was stepping into was free, uncorrupted, and peaceful. George hoped so, too.

There was action on that trip, but it wasn't the kind George had initially worried about. Ten Canadian pilots were assigned responsibility for ferrying Bell 212 Helicopters from Mogadishu to Mozambique, through a route that had them first overnighting in Mombasa, Kenya, and then making a stop in Tanzania before completing their mission.

They were at their stop in Mombasa, Kenya, when disaster struck: one of the pilots hovered too close to a small airplane, causing the fixed wing to flip on its side. The aircraft was wrecked by one small, thoughtless act.

Air traffic control insisted all the Canadians return to the airstrip, and they grounded the pilot who had caused the accident. There were many arguments and much confusion between the ATC, the owner of the plane, the helicopter pilot, and the company George and the other Canadian pilots worked for, but eventually it was decided that the nine pilots who had not caused a crash could move on, and that the guilty party's chopper would stay behind as collateral until the company he worked for paid for the plane's repairs. This meant leaving the pilot behind, too.

It was an embarrassing situation for everyone, and hard not to feel guilty leaving a man behind. Later, George would learn the Canadian company had taken weeks to pay. When they finally did, George was transported back to retrieve the chopper and fly it to Johannesburg, South Africa.

But that was the worst thing to happen on that assignment and, after the Rwandan genocide, Mozambique was a great reprieve. George was stationed in Quelimane, a seaport city near the mouth of the Rio do Bons Sinais, the "River of Good Signs." The election was peaceful but blistering hot, and the people of Mozambique lined up their shoes and hats under the hot sun to save their places in line while they themselves waited under the shade of nearby trees and buildings.

The Front for the Liberation of Mozambique army (FRELIMO) had agreed to put their weapons down following peace talks with the United Nations and, largely, George enjoyed the assignment. He found pleasure in seeing the little children who would sit in a straight line under the shadow of his helicopter rotor blade, thinking it funny and convenient that its blades cast shade on the ground just big enough to shelter their small bodies. He almost started World War Three handing out assorted licorice to the children who greeted him, Debbie always packing extra candy in his bag so that he might share. He bought a small bicycle in Somalia and rode down to the ocean shore each day after work, leaving the bike to a hotel worker who had befriended George during his visit. In that case, George had to write a letter that proved he consented to give the man the bike, as the vehicles were such hot commodities the worker feared police would assume he had stolen it. That man wouldn't have stolen a thing, George knew. The only theft George experienced on the whole trip was the loss of an electric fan, snatched by the United Nation's Russian contractors!

When the assignment officially came to an end, Samora

Machel was named the first president of Mozambique, and his wife, Graca Machel, would become the first woman in history to be the First Lady of two different countries, later marrying South Africa's Nelson Mandela.

George was glad to see the country had survived such a momentous moment in its history without bloodshed. Mozambique was not Zimbabwe, but it was close enough to the home George once knew that being on assignment there felt something like visiting an old friend and, indeed, their peace and good fortune was more like the country George had grown up in than the country Rhodesia had become, for Zimbabwe had fallen into extreme poverty with inflation so high that trillion-dollar bills were printed and easily obtained. What had once been the breadbasket of Africa was now in such neglect that small children lived with stomachs distended from hunger. George hoped Mozambique would not fall in such a way and was momentarily happy for the nation. Later, Samora Machel would be killed in his plane, but at that moment, the country celebrated. Mozambique had been reborn.

His work complete, George was next assigned to work with the Italian explosives team, transporting the soldiers and equipment necessary to rid the area of hidden landmines and weapons left over from civil war. The idea was that the United Nations would buy weapons off guerilla organizations, encouraging rebel soldiers to turn over landmines and caches of armory hidden in the jungle. The United Nations would then send the Italian explosives team into the jungle with George's help of transport, where they would use dynamite to blow up massive collections.

Flying to Maputo in Mozambique to rest and refuel before completing the second part of his journey, George found himself flying just 175 miles from Umtali and recognized the landscape of Beira, below. He was flying over The Estero; the Grand Hotel where John Leonard and Louise Susannah used to take little George and Marion to holiday, the same one that wouldn't allow peasants like the Kelhams to darken its elite threshold. George looked down on the massive building, now turned relic. The hotel was a shadow of what it had been thirty years earlier. It was laid waste; all windows broken, doors removed, walls sledgehammered through. Garbage filled the crumbling swimming pool. Abandoned by its owners, the hotel had been overtaken by hundreds of squatters and had turned into a sort of over-crammed city of its own: one that was violent and dark, reliant on black market dealings that included the terrible trafficking of drugs, sex, and people.

In some ways. George was sad for the old hotel. Its demise was such a waste. But systems were shifting. The outdated institution had relied on elitism and entitlement to preserve its fortune, its standing, and its success. The same systems that would have preserved that hotel would not have seen George take to the sky above it.

Good riddance to old ways, George thought.

Success was no longer a passive show of opulent, meaningless wealth.

Success, for George, had come with the ability to keep moving, and keep striving, doing his best. It was a thing to be earned, and earned again, daily.

~

When George's work with the Italian bomb detonators ended, he returned to Slave Lake and started driving a log truck for Bob Vanderwell, who owned a local lumber mill. It was minimum wage work, and George was lowest on the totem pole again. But Bob had taken a liking to George. A deeply religious man, Bob appreciated that George didn't drink or use drugs, and could be counted on to show up on time for work every day, or even cover another driver's route when they failed to. George didn't know it at the time, but he was receiving a master's education in business just by knowing Bob Vanderwell.

George was put on the nightshift run at Bob's mill. His work ended around four AM, but it wasn't unusual for Bob to be on site at that oddball hour, and then the men would chat. At all hours of the day and night, Bob Vanderwell would show up to the mill, investigating even minor issues himself, keeping abreast of all aspects of his multi-million-dollar business. It was impressive. He didn't have to be so hands-on, after all; Bob could easily have afforded to have paid someone else to manage the mill. But Bob cared more about his business than he did his pleasure. Already massively wealthy, Bob had no need for extra money, but he knew that the many local workers he employed sure had use for their jobs. Bob kept the mill humming along like a well-oiled machine because he recognized the stakes of his business were higher than his own fortune.

One day, in the very early hours of the morning, George was just about to finish his night shift. Bob Vanderwell happened to be on the floor, and George asked if he might wash his pick-up truck in the mill's heated washing station. He knew Bob to be friendly and giving, after all.

"Absolutely not," Bob told George, surprising him. "I employ almost two hundred people, George, and if anyone sees you washing your truck here, then they'll want to wash theirs', too. Either my bay will fill up with people taking care of their vehicles when they ought to be working or when my own trucks need in, or feelings will be hurt, and folks will think I play favoritism. No. Work is work. You must keep your personal life separate."

Without realizing it, George had started logging away the lessons he was learning from Bob.

Stay involved.

Business is business.

Know the people who work for you.

George didn't know why he had started listing these points of philosophy in his head, but he had an inkling. He imagined how exciting it might be to someday own his own business. While the thought was only in its conception, it shone in his mind like a beacon, drawing him home.

In the summer of the same year, George was brought back to the United Nations for one last six-month assignment. He would be stationed in Guatemala and charged with transporting peacekeepers, political leaders, and, occasionally, medical evacuations.

The Guatemalan war was terrible, but violence, at that time, seem contained to more rural areas. Still, within weeks of being in the city, George and his helicopter were featured on the front page of the Guatemala Gazette alongside corpses covered by body bags. His craft flew the massacred people from a mass grave so they could receive proper identification and burial. Cities, however, were patrolled by heavily armed guards, so George felt they were still safe enough to bring his family to visit.

He paid for a fine hotel room, arranged sightseeing tours, and invited Debbie and the children to visit. While he flew negotiators through the rainforest, the other three Kelhams would enjoy themselves before he returned to them in the evenings.

When the boys arrived in Guatemala, George was taken aback by how big Boyd and Taylor had become in the short time he'd been on mission. George saw them every second month, but it was as if they saved up all their growing for when he wasn't around. They grew not at all, and then all at once, like sunflowers stretching to meet a bright day, or grass after a heavy rainfall. In Slave Lake, the boys played hockey and had gangs of friends that met at the park behind the Kelham house. They raced dirt bikes and, just like George, they both loved music. And while George was wholly grateful for the ability to provide for them through his pilot's license, part of him worried he was missing an important part of their lives.

But it wasn't until Taylor, only nine years old, asked about the bullet-proof vests that lined the floor of George's chopper that George realized that they missed him, too.

"Oh, they're only there because they help block out the sun," George fibbed to his little boy. "Don't you worry about your dad."

Taylor, small but smart, squinted his eyes. He wasn't convinced.

"You should be careful, Dad."

With that, George knew it was time to rebuild a life at home.

PART FOUR

BRAVO HELICOPTERS
1995

The level of commitment George had for his family could be demonstrated in a singular act. Adamant to work closer to home, George Kelham reached out to Far Away Helicopters and asked Willy Light to bury the hatchet.

It was the same snot-nosed crony who'd fired George that took his call and later met George at a local truck stop. Willy's refusal to meet with George was an obvious sign that he wouldn't be rehired, but, again, Willy's friend didn't have the guts to say that to George's face. Instead, George was told Willy would get back to George about the proposition later. Of course, he never did.

There was no flying work to be had in Alberta, but George was able to land a decent-paying job with Campbell Helicopters and the company kept him busy flying over Ontario, Idaho, and California. On one hand, he was working over the same continent his family lived on. On the other hand, North America was so vast it really didn't matter whether he was in Canada or Timbuktu.

George worked for Campbell for about six months, flying home to his family every three weeks for a few days of downtime. When that work ended in the winter, George once again returned to Bob Vanderwell's employment.

As good an employer as Bob was, driving truck wasn't what George was meant to do. He was qualified for so much more, and far more money could be made flying a chopper. George and his family hadn't sacrificed so much of his life for him to spend what remained of it on the road. In the spring of 1996, George resumed his mission to find local flying work. He contacted five companies, but no one was hiring. Finally, George decided it would be up to him to create work for himself. If there wasn't a local employer hiring, George would bring a non-local employer to his ideal location.

George knew it would mean going toe to toe with Far Away Helicopters, and, yet again, working his ass off for another man's business, but that didn't matter. George promised himself he'd be closer to his family, and George was a man who kept his promises.

George had heard of Bravo Helicopters[13] many times and knew Leroy Dean would have insight into the company. George called his old friend up and asked what he knew. Leroy said he himself wouldn't be interested in working under the owner, Dan Smith[14], but that Bravo might be interested in a Slave Lake base. Smith, Leroy said, seemed to be always interested in a worthwhile expansion.

George found Dan Smith's number and called him up.

"Give me a helicopter and a cell phone and I'll make you a lot of money," George said, cold calling the man.

On the other end of the line, Dan laughed a hearty laugh. He asked if George was available to meet and George found himself driving to St. Albert only moments later. There, Dan treated George to lunch at an Italian restaurant named Sorrentino's.

"Okay," Dan said, over breadsticks. "I like your style. Let's give this a try."

George didn't know it then, but at that very moment he had created the third of what would become his major competitors.

Life with Bravo started off well. Bob Vanderwell transferred his tree-planting account over from Far Away Helicopters to Bravo in support of George. Buchannan Lumber – the owner a friend of Bob's – was also ready to make a change. Almost immediately, George brought in so much work that Debbie had to help manage Bravo's bookings and perform administrative tasks. Dan Smith was impressed in his new base manager as well as his parttime secretary, even though he refused to add Debbie to payroll or so much as send her a bouquet of flowers as thanks

[13] Company name fictionalized.
[14] Name fictionalized.

for her hard work. George was excited by how quickly the business was becoming established, but most of all, he loved pissing off Willy Light, who was furious to find himself competing with a man he had once employed.

"You know, Willy should have known better," George told Debbie as they celebrated Bravo's good fortune. "Look at how a mere hospitality manager situated his own hotel as the Sawridge Hotel's greatest competition in town. Willy knew the Sawridge should have paid Bert not to go into business elsewhere. All a skilled worker needs is a little bit of money behind him. Did Willy think I'd never get a little bit of support behind me? He could have given me a hundred grand and had me sign a non-compete contract, and he never would have seen me again. He'd own the whole sky above Slave Lake. But Willy was arrogant. Now I'd fly for free if it was the only way to prevent him from taking an account!"

Within a few short months, George had brought in more and more contracts, the major ones being CNRL and Alberta Plywood. George created so much work that Dan had to send more helicopters to Slave Lake.

The trouble was, Dan could be a chaotic supervisor. He'd need George up in the sky at a moment's notice, making family plans impossible to orchestrate. George would miss backyard barbeques with friends, date nights with Debbie, outings with the boys.

Losing these moments of his children's lives could sometimes be infuriating, but George and his family weren't the only ones who suffered due to Bravo's disorganization. Dan's mismanagement often cost his own company money, too. One morning, George was having a coffee with friends when his flip phone rang. Dan asked George to head to Ontario, where folks were waiting for George to assist with a wildfire. Dan's instructions were to start flying immediately, and to phone him at every second fuel stop until George arrived, as back then there was no GPS to track the trip. So, George called Debbie and asked her to pack a bag for him. He grabbed his luggage, stopped by Bravo's hanger, and jumped into a 206.

George made it to Saskatoon before stopping to refuel and call Dan.

"Where are you at?" Dan asked.

George was confused. "Saskatoon. En route to Ontario,

right?"

"Okay," said Dan. "Stay in Saskatoon. Grab a cab and go to a hotel. I'll call you in the morning."

"Don't you want me in Ontario?" George asked.

"Just hang tight for now."

George didn't mind the change of plans one bit. It wasn't his money lost if Dan didn't secure the wildfire account. With nothing else to do, George made a minivacation of the trip. He headed downtown into "The Paris of the Prairies" and enjoyed a steak dinner before buying a ticket to watch *ConAir* at a movie theatre. George returned to the hotel around eleven PM, and enjoyed a long, deep, lovely sleep on crisp hotel sheets.

Always an early riser, George had finished reading the paper and sipping coffee before his phone finally rang at seven AM.

"Where are you?" A voice shouted on the other end of the line. It was Dan's son, Peter[15].

"Saskatchewan," George replied.

"Saskatchewan? What the hell are you doing in Saskatchewan? Get back to Slave Lake! You can't run a base from there!"

After hanging up the phone, George took a deep sigh, folded up his paper, and flew home.

It turned out the wildfire contract had been wishful thinking on Dan's part. George could have told him that. A pilot can't rely on forestry fires to pay the bills. Chasing fire was like trying to catch lightening in a bottle – Forestry didn't care who they gave fire accounts too, they would always choose the closest chopper.

But that wasn't a big deal to Dan. With George on his team, Smith was pulling in buckets of cash.

But despite tripling Bravo Slave Lake Base's annual revenue, George wasn't making any more than his usual, humble, wage. By October 1997, George insisted Dan acknowledge the work George had performed.

"I've shown you what I can do," George said. "Now I need something more, Dan. I want a helicopter and a subcontract. Or I want to buy the hanger from you and become a partner. What can I do to be taken seriously?"

Dan shook his head. "This is a family business, George. I can't take you on as a partner. My kids are going to inherit Bravo."

George wanted to talk more, but Dan requested to leave it to

[15] Name fictionalized.

another time. So, George waited, and then asked again in a few weeks. Again, Dan refused to speak. Soon he was giving George a run around so brisk that George felt like he was in a game of cat and mouse.

Unable to get his employer on the phone, George knew his time with Bravo was coming to an end.

He went to Debbie for advice. Like always, he was right to trust her for her grounded, tempered insight, and despite how sick he felt about the situation, she made him feel better.

"If he won't make you a partner," Debbie said, "see at least if you can get a raise." She smiled and shrugged like it wasn't a big deal whatever happened, but George could see in her eyes that she was scared. She didn't want to lose him to far-flung work again, even if she would never try and stop him from chasing his dreams or demanding his worth.

About a month had gone by since the first meeting, and George found he was kept up at night by the nagging, unrelenting truth that something wasn't right, that he was being taken for a fool. Finally, he called up Dan and insisted to meet.

"Dan," he said, "What I have to talk to you about cannot be discussed over the phone. Either you are coming to Slave Lake, or I am coming to St. Albert. What will it be?"

Dan sounded shocked. His voice raised an octave.

"Well, uh, ah… I need to be here in the morning, George. And you can't leave the Slave Lake base unattended!"

"Fair enough," George said. "So, we'll meet halfway, in the evening. I'll see you at the truck stop in Fawcett."

Dan didn't like it, but reluctantly he agreed.

They sat opposite each other in the cigarette smoke filled café. There were gas pumps outside the window, washrooms and a gift shop to the left. Dan sat scowling. We're not in Sorrentino's anymore, George thought to himself.

"So, why'd you drag me here, George?"

George took a breath. "Dan, I've shown you what I can do for you," he said. "Now show me what you can do for me."

Dan shrugged. "I have no idea what you mean."

George's heart sank. Already he could feel the conversation wasn't going to end well.

"I brought in a huge amount of money this summer," George said. Dan wouldn't acknowledge the number, and George felt flabbergasted that the man was in such denial. "I've tracked the numbers, Dan. I'm not a fool. You're paying me sixty thousand a year, and it just isn't cutting it. You've got to give me something more."

Dan shook his head. "I told you, George, this is a family affair."

"Then more money," George said. "Give me a raise."

"How much?"

"I want a hundred thousand a year."

Dan laughed. "No one's worth that much."

"Then give me ten percent of whatever I bring in," George said. "Just wait and see how much work I'll bring you!"

Dan shook his head. "No way. Not a chance."

"Well, make me an offer then!" George almost shouted, his hands coming down hard on the table in frustration. He felt like the room might start spinning, he was angry and heartsick and nauseous. It took him right back to the argument he'd had years earlier with Willy Light. George was tired of making thoughtless men rich.

Dan was silent for a moment.

"Listen," he finally said. "I'm going hunting in Fort Nelson in November. I'll mull this over then, and when I drive through Slave Lake, I'll have a number for you."

Anger rose in George. His face was burning.

"No," he said. "I promised myself and my wife that we would make a deal tonight. If not, then I must move forward without Bravo."

"Please," Dan said. "Trust me, George. What other option do you have?"

Angry at Dan, angry at himself, George took Dan at his word and left the truck stop.

A month later Dan stopped by the hanger on a bitterly cold November day. There was an apprentice pilot in the building washing choppers, so George and Dan sat in Dan's truck to hash out the details in private.

"George," Dan said, "I'm going to give you a raise."

George was so delighted he almost clapped. He had expected the worst, but now felt as though he would erupt in joy. He couldn't wait to tell Debbie it was all going to be alright.

Dan smiled, pleased that George was happy.

"Five hundred more a month," Dan said, "and you can keep my credit card to fill your truck with gas. And every Christmas, I'll give you a hundred dollars for every month of the year you worked."

George's face fell.

"Woah, woah, woah," George said. "That's not going to work for me. That's no good. We are way too far apart here."

Dan narrowed his eyes. "Well George, that's what we think you're worth."

"Dan, you don't have Vanderwell's without me." George said. "You don't have CNRL, the Buchannan's, or Alberta Plywood."

Dan laughed. "Well, we'll see. What are they going to do? Go back to Far Away?"

For four weeks, George barely slept for he was so sick with rage and resentment. Debbie was concerned. Finally, one night after the kids had gone to bed, George turned to Debbie.

"Dear, I think we have to shit or get off the pot. We can either enjoy what we have or go out on our own."

Debbie smiled. "George, what took you so long? You do what you need to do to make this work."

~

The first thing George did was talk to Bob Vanderwell. Years ago, Bob had asked George why he was working for other men instead of starting his own business, but George had laughed him off, believing that perhaps Bob had amassed too much wealth to remember what it was like to be the little guy with no money.

"Remember that idea about me starting a helicopter business?" George asked. "Well, it's going to happen. And I might need your help."

To George's surprise, Bob said he'd do more than help. He wanted in. Bob Vanderwell proposed they form a partnership leaning on George's skills and Bob's money. George was interested and told Bob he'd have to talk it over with Debbie.

But to George's even greater surprise, Debbie refused.

"This business will be ours or it won't happen," she insisted. "No partners."

George was stunned. She knew buying a helicopter could cost them every penny they'd ever saved and much more, and her refusal might have been the first time she'd ever said no to any of George's career plans. George was intimidated by the idea of going into business alone, but he was impressed by his wife's courage. They decided that instead of relying on Bob Vanderwell, they would take the money from their home equity.

"Okay, Debs," George nodded, filled with admiration. "Just you and me."

Bob took the rejection well. Business is business, after all – Bob himself was the one to teach George that. He offered to write George a letter of endorsement instead, noting for the bank that he was committed to being George's first client once Slave Lake Helicopters was ready to open shop. That was such a good idea, George realized, that he asked George Gross of Buchanan Lumber to do the same, as well. Mr. Gross was more than happy to comply; he knew the great service his company had enjoyed had nothing to do with Bravo, and everything to do with the pilot he'd flown with.

George and Debbie took both letters of guarantee to the Royal Bank, along with the title to their house. The Kelhams had paid off their mortgage, but they agreed to pull no punches – they would take whatever risk they had to take.

"We can give you a $75,000 line of credit on the house," their banker said. The banker was the wife of a pilot George knew.

"How the hell am I supposed to buy a helicopter with $75,000?" George asked.

The banker just shrugged.

It was impossible, George thought. Then again, he'd made the impossible work before.

"We'll lease a chopper," George told Debbie. "It's the only way."

George made an appointment at Avialta Helicopters, a helicopter repair and overhaul service center that also specialized in machine leases and sales. George knew the owner, Rod Wood, to be a good man. Before his arrival, George told Rod the whole operation was top secret.

"I mean it," George had warned. The stakes were high. If Dan Smith realized George was considering going into business for himself, he'd lose his job and the Kelhams would quickly go into too much debt to use the home equity for a chopper. "You can't say a word to anyone."

Rod took George to the back of the hanger and showed him a Bell 206, but it was far too small for the work George hoped to secure. Next, Rod showed him an AS-350 B2, but it was far too expensive. George was starting to feel like Goldilocks in the story of the three bears. Nothing was right. The last chopper Rod showed George was a "longdog" – a Bell Long Ranger, Registration CGGSL.

It wasn't the helicopter of choice for planting trees, but it was large enough, and it was the right price.

"Well, I guess I'll take the long dog," George said, stifling a groan.

Back at the head office, Rod's manager helped George fill out the necessary paperwork to lease the helicopter.

"Now, there's quite a bit of information we need before we can finalize this," the man warned.

"Of course." George said. "But how much is the deposit?"

The manager reported that the cost would be a ten-thousand-dollar damage deposit, plus the first months' lease, and insurance. George felt his heart race. He had a total of $75,000 to get this business off the ground, and in only the first month he'd have to spend $43,000. He forced himself to breathe slowly, to not let his nervousness show. If this idea failed, it would take quite some time to pay that much money back - especially considering George would be unemployed!

"It's not a problem," George said. "I will get this information in order within the next week or two. But," George warned, again, "This needs to stay between us. If this gets back to Dan, the deal's dead."

The manager extended his hand and shook George's. "You have my word."

Of course, the plan hinged on far more than just one man's discretion. The last thing George had to do was gain approval from Transport Canada. The organization would first need to certify George to work as a Helicopter Business Operator before he could even apply for insurance. But Dan Smith had close friends within the organization. All George and Debbie could do was hope professionalism would prevent the workers at Transport Canada from leaking the news.

The Kelhams were sick over it for days. Finally, George got the call: on January 26th, 1998, George would be issued an operator's license by Transport Canada. George absorbed the news with the same quiet delight he'd felt when doctors had told he and Debbie when their youngest child would enter the world.

"That's Wayne Gretzky's birthday, Dad!" Taylor had exclaimed over dinner when George and Debbie told the boys about the change in their parents' careers.

George laughed. "I'm sure that's a good sign, then!"

"I get to be company president!" Taylor shouted.

"Then I'll be CEO," Boyd said.

Taylor narrowed his eyes and looked at his dad. "Is CEO higher than president?" he whispered.

Boyd, now in his early teens, was amazed his parents could do something so big.

"So, like, we're going to *own* a helicopter?" He asked, suddenly serious.

"Well," Debbie corrected, "We're going to *lease* a helicopter, and own a helicopter business."

Boyd didn't care. He couldn't get over the fact that his seemingly normal, working-class parents were going to have such a luxury item in their name. The boy's eyes twinkled, and he smirked.

"So, like, is this really *taking off?*"

George exploded in laughter. He reached over the table and messed his son's hair.

"Yup," George said. "That's the plan. We've just got to *take off.*"

~

The only thing left to do was to tell Dan that George was leaving.

On the morning of January 26th, George picked up a group of diamond miners in Bravo's red-branded helicopter. Something of a diamond rush had sprung in northern Alberta, and more and more, groups of people were hiring George to deliver them to remote locations as they searched for kimberlite.

"When you're ready to fly tomorrow," he told the group, "There will be a blue helicopter parked across the road. Drive over to the blue one, not this red one. Tomorrow, your helicopter will say Slave Lake Helicopters on the side."

The passengers just shrugged. They didn't care.

"Are you still going to be the pilot?" The leader asked George.

"Yes."

"Well, that's all that matters to me."

After dropping the passengers off, George flew back to Bravo's Slave Lake base and cleared out his personal belongings.

He gathered the few things he owed Dan – the keys to the hanger, his credit card for gasoline – and drove down to St. Albert, climbing the narrow stairs to Bravo's office. Dan wasn't in the building, but his son, Peter, was.

"I'm leaving," George told him, handing over Dan's things. "Here's your stuff, and everything else is in the hanger. It's all intact."

"What are you talking about?" Peter asked.

"I told Dan that I need more money for the amount of work I'm bringing in for you. I'm not being fairly compensated."

"Woah, woah, woah," Peter said. "Slow down a minute. I'll sort this all out."

George laughed.

"Too late for that," he said. "I've got a machine waiting for me just across the road at Avialta Helicopters. I'm going out on my own."

Peter snorted. "Oh yeah? Whose Operator's Certificate are you going to use?"

George smiled. "My own," he said, walking away. He took pleasure in the sight of Peter fuming, uncomfortable in the truth that Transport Canada had not given Bravo full warning.

As George walked across the shop floor, the door opened. Dan entered the building.

"What are you doing here?" Dan asked.

George sighed, annoyed to repeat himself.

"I've just explained it to your son. I'm going out on my own. Starting my own business. I gave everything I owe you to Peter, but I've got to get going." It was winter, and there weren't much daylight hours left. George had to get home before dark.

"No, you're not," Dan said. "Come into my office. We'll have a coffee."

George laughed incredulously. "I don't have time, Dan! I have to get home."

For the first time since George had met him, Dan was speechless. He just stood there, open mouthed and confused, to the point that George wasn't sure what to do. Finally, George shrugged and continued across the shop, feeling Dan's eyes follow him out the door. Then he walked across the road, got into his helicopter, and flew home.

At the Slave Lake tarmac, George's chest filled with pride to see his little family waiting for him, waving, and smiling from the ground. All four Kelhams witnessed Slave Lake Helicopter's first flight and, the next morning, George flew his first customers – the Kimberlite prospectors – happily, to their site.

SLAVE LAKE HELICOPTERS
1998

Before picking up the keys to the helicopter, George and Debbie had wondered if January was the right time to open their doors. Winter seemed the worst time to start a helicopter company; there was little work, and northern Alberta's brutal climate kept many machines grounded. They played with the idea of putting off starting the business until spring, but George just couldn't keep working for Dan Smith for that long. Every hour he spent building Bravo's business, he was stealing from his own.

Fortunately, the winter of 1998 was mild, and while there wasn't much traditional chopper work to be done, there was the unexpected kimberlite and diamond rush. Soon George was flying prospectors all around Alberta. Debbie couldn't believe their luck.

"Someone up there is watching over us," Debbie said.

"My mother," George said aloud. He still wasn't much for religion, but in that moment, he felt his mother's protective love as powerfully as he did on the day he left Umtali via rail to the military training base.

Debbie smiled. "And mine," she said. "And Mrs. Mayo, too."

How lucky they were to have the eternal love of such wonderful women.

They set up an office in a tiny rental space within the terminal building of Slave Lake Airport. It was so small that if the couple had a client visit them, George had to stand half in the hall and half in the doorway as Debbie took notes from a small desk and the visitor sat opposite in the lone chair.

The couple soon discovered Debbie's experience assisting in administration for Bravo had paid off. She was a natural in the seat of command: booking the helicopter, invoicing customers, filing relevant information securely away. Even more fortunate was the friendship Debbie continued to share with her former employer, Marilyn Dawson. Marilyn was always only a phone call away, happy to help with any administrative questions. Just as Debbie had when she and George first moved in together, Debbie set to work making the office as organized, homey, and welcoming as possible. The long dog was George's domain, the office was Debbie's and, together, they were full partners.

By late February, however, northern Canada's excitement over diamonds had died down. Days went by one long, slow minute at a time, George having washed the helicopter to the point where it shone, Debbie checking the phone again and again to make sure it wasn't broken. The Kelhams were a bundle of nerves. When George thought about all the money they owed, cold chills ran through him. Their family home, their boys' futures, and everything they'd ever achieved was tied up in a great big expensive risky business.

"You did your homework," Debbie reminded him when he spoke of his worries. "It'll pick up."

George nodded. She was right. He had spring contracts to move saplings and tree planters for both Vanderwell Mill and Buchanan Lumber – Slave Lake Helicopters just had to hold on until planting season. Besides, Slave Lake Helicopters was local. That meant they'd be one of Forestry's first choices for transporting wildland firefighters.

But chasing fire is like trying to catch lightening in a bottle, George's mind would remind him.

He did his best to ignore the thoughts and made more calls, instead.

Again, it was vital for George to find a way to make his career work in the growing town of Slave Lake. Debbie and the boys had made a lifelong home in the area, and George couldn't imagine asking them to leave it. Debbie had created important friendships and bowled on the ladies' team every week. Boyd and Taylor had grown into polite, good-natured kids with many friends, and seemed loved by their teachers and the employers who provided the boys after school jobs.

Yet, as valued as the boys seemed to be, conversations of their school and work often turned to minor arguments between father and sons. Boyd didn't enjoy school except for the friends he had there, and George would become frustrated at his teenage son's poor grades and lack of studying. By the time Taylor entered junior high school, all of George's arguments over scholarship had already been had with Boyd, but he then found himself critical of work ethic. If Taylor's dishwashing shifts started at six PM, why didn't he begin at five PM to show his initiative? The boys would only shake their heads and walk away from these conversations. George knew he came from a different world than his children, and that perhaps he was pushing them more than necessary, but he only wanted them to succeed. In his experience, life was not as easy as the comforts of Canada often made it seem. It was an irony George wouldn't realize until much later: the things he and his sons fought over most were the same subjects that created distance between George and his own father.

Still, George beamed with pride whenever he ran into anyone who knew a member of his family, which was very often as Slave Lake was still a small town. His wife and children were loved and supported, and their friends seemed to wish nothing but success for Slave Lake Helicopters.

But others in the community were not so positive.

One day, George and Debbie were driving around on a miserable, slow day, trying to cheer themselves up with a couple of ice cream cones and a change of scenery. There, on the side of the road, George saw a couple of men he'd worked with at Far Away Helicopters. He pulled over to say hello through Debbie's passenger side window.

"So, you think you're going to take on Far Away, eh?" One of the guys laughed.

George shrugged. What was it to him?

"I'll certainly try my best," George said. He could feel Debbie tensing beside him.

"Willy's going to *destroy* you!" The other man exclaimed. "Oh man, he's gonna run your ass out of town!"

Debbie turned to George, aghast. George lifted his hand and bade the men farewell. He simply rolled up the window, put the truck in drive, and kept on. Debbie was irate but George didn't want to talk about it. Things would take off, he'd promised himself. That was the plan, so that's what he'd make happen – no matter what.

It was reported later to George that, after Vanderwell pulled out of Far Away's roster, Willy Light had gone with a friend to convince Bob to come back.

"We'll offer you better rates," Willy had promised.

Bob only laughed.

"Where were the good rates before George showed up?" he asked, showing the men the door.

Spurred on by pride and a competitive nature, George threw his salesmanship abilities into overdrive, cold calling everyone he knew and even more that he didn't. Slowly, business picked up. Debbie and George bent over backwards for their clients, so eager to show their appreciation, so wanting to be known for excellent customer service. The months moved on at a grueling pace, and Debbie kept a constant eye on the dollars that drained from their bank account. George tried not to think about the risk of losing the equity in their home.

Finally, spring came. With it, a large forest fire started outside of Chisholm, a small hamlet about fifty kilometers outside of town.

Slave Lake Helicopters had the closest machine.

Suddenly, time sped up. Now the problem wasn't passing slow minutes, but in filling three tasks where usually only one would fit! George began his days as early as four AM, leaving Debbie asleep in their bed. He'd get the machine warmed up and ready to begin work at first light. Debbie would wake a few hours later, tidy the house, get the kids to school, and get herself ready for work. Then she'd meet George at the office to take care of the books and set up new work. She kept George in the air from sunup until sundown, and George loved it.

At dusk, he'd set his helicopter down for the night next to the terminal building, and then he would fill large buckets with warm, soapy water, leaving the pails beside the machine. George would then drive home, and Debbie would kiss him hello before she left to wash the chopper and put it to bed. George would stumble into their kitchen like a zombie and eat a few bites of whatever food Debbie had left him on the counter, then he'd send the kids to bed before climbing the stairs and falling into immediate sleep himself. For her part, it was as though a phone had been grafted to Debbie's hand. She would receive calls from clients as late as three AM and still jump to attention, booking helicopters, taking down names and dates.

The next day, the alarm clock would sound at four AM, and the same routine would happen all over again.

There were no Transport Canada limits on flying hours in those days, but there were physical limits, and emotional limits, too. For all of George's want to work close to his family, it felt like he never saw his sons. Both the Kelham boys had done well for themselves in hockey and motor cross, but George was rarely able to see them play or race. On the same day Boyd's hockey team won the Alberta Peewee playoffs for the first time, George missed the occasion, having had to fly over thick bush at last minute.

Sometimes over-exhaustion would cause him to snap at the boys. Hoping to protect them, George insisted they rest before big games or races, but he was not home to enforce the rule. One day, Boyd sneaked in a few practice laps the morning before a dirt biking competition and he wiped out, breaking about a million little bones in his wrist, causing him to have to withdraw from the tournament. Another time, Taylor had taken the same risk and had crashed, writing off his machine just hours before he was scheduled to compete. George had been so angry over the phone, so concerned for his sons but also so frustrated that he, away flying, could do nothing to help.

But other times George found ways to be there, even when he was not. On days when he couldn't leave his helicopter, Debbie would call ten minutes before the boys were about to begin a race and he would time his patrols so that he could hover over the boys' tournaments, cheering down onto racetracks as loud as he could despite no one being able to hear his voice above the noisy machine.

"Yes!" George would shout into the noise. "Drive! DRIVE!"

While the Kelhams were thrilled to see their business flourishing, George's health was starting to fray from lack of rest. He was juggling too much. The companies he was working for turned a blind eye to his practice of double booking: picking up the forestry fire boss in the morning for meetings before picking up tree planters while the meetings took place, and then grabbing enough saplings to keep the planters busy for six hours before going back to the office to pick up the fire chief and deliver him home for supper before returning to the tree planting site. Slave Lake Helicopter's clients wanted the company to succeed so they didn't mind that George was booking out their time to others, so long as he didn't keep them waiting. George didn't keep them waiting, but he couldn't keep up that speed forever, either.

Eventually, George realized he didn't have to. He started to look for help.

The Kelhams bought a helicopter – a Bell 206 on a lease-to-own contract – and started interviewing employees. They subcontracted pilots: first, an older pilot from British Columbia, then a pilot named Roger, and another named Stan. All were fine men, but their contracts were short lived as rural Slave Lake was such a difficult place to attract skilled workers to. It wasn't until Leroy divulged that he was at a crossroads in his career that Debbie and George finally found the perfect fit.

"Would you come work for me?" George asked Leroy.

Leroy just laughed. Despite his consideration of retirement, he was delighted by his friend's success.

"Yeah," he said, nodding. "Of course I'll fly with you, George."

Leroy was skilled and experienced and, best of all, George and Debbie trusted him. But while his presence took stress from George's shoulders, the company was enjoying exponential growth and tasks weren't slowing down. Another pilot had to be brought on, but George didn't know any available for such work.

Finally, the strain came to a breaking point. George was asked to transport firefighters for a large contract, but he and Leroy were booked solid the next day and unable to accept the opportunity. George needed a pilot immediately. He'd have to take a risk and hire someone he'd never met before.

But George had heard good things about a female pilot named Andrea. She was said to be friendly and quick as a whip – intelligent and efficient. She flew throughout Alberta but hailed from Quebec, where her husband and children lived. She was a francophone by birth, but George soon learned she spoke English with the eloquence of a poet and was a true philosopher at heart. Andrea was in town to fly for another company, but the employer had failed to secure the proper documents to fly over wildfire, so all their machines were being called home.

George met Andrea at the hotel room she was staying in and knocked on the door.

"Hello, Miss," George said when Andrea opened the door. "I will not come in, not because I want to be rude but because I don't want to put you in an uncomfortable position."

Andrea pursed her lips to contain a smile. She was a woman who'd been flying in a male-dominated field for many years and had dealt with men much cruder and ruder than the gentleman standing at her door, now.

"How can I help you, Mr. Kelham?"

"I've heard you are a good pilot," George said. "I have clients who need to be transported tomorrow. I don't have time to do it. Would you like to pick up some extra work under my operational license as a subcontractor?"

"Sure," Andrea said.

George had warned Andrea that his clients expected him, and that they most likely would be disappointed to see her. She told him not to worry and insisted the flight would go well.

The next day, George's clients called Debbie to let them know how much they enjoyed the company of Slave Lake Helicopter's new pilot, the charming and beautiful French woman. After that, George and Debbie invited Andrea to work for them fulltime.

Life became manageable, then. George and Debbie had the support of family and dear people whom they'd first hired as employees but later treasured as friends. Andrea and Leroy had supper almost every night with Debbie and George, and George threw his energy at parenting with as much gusto as he had in starting the business. Staying true to the promise he made to himself as a young man, he would spring the kids from school to take them to whatever concert he wanted to see, just as he had taken Taylor to see The Rolling Stones those many years prior. ACDC, Def Leopard, Motley Crew, Metallica, Black Sabbath – music was a major bonding activity for father and sons. He bought season tickets to see the Oilers and took Debs for date nights to Edmonton a couple of times each month.

The family gave back to their community in a larger way, too. George joined the rotary club and the Kelhams took in a foreign exchange student named Charlie, a boy from Thailand who lived with the Kelhams for a year and quickly established himself as a favorite in the household. Charlie was bright and funny, patient and kind. He immediately fell in with Boyd and Taylor, and George and Debbie treated the boy like their very own, delighting in his curiosity and amazement at Canada's land and culture, just as George had done when he had first arrived. The Kelhams spoke with Charlie's parents over the phone, keeping them abreast of their son's adventures and wellness.

"You will have to visit us after Charlie comes home!" Charlie's father insisted. "Promise me you will let us host you!"

"Absolutely," George agreed. "We would love to meet you."

Soon, Slave Lake Helicopters had earned enough money to put a deposit on a third machine – this time, an A Star.

In the years since George had insisted Willy Light purchase an A Star, the machines had remained by far the best option for carrying out the work a northern Albertan helicopter company had to do. Still, even a used chopper costs over one million dollars. When the money came through and the time came to put it all down on the A Star, George looked at Debbie with nerves tying a knot in his stomach. Were they doing the right thing? George thought of his childhood and of the tiny, square house he'd grown up in, and of the third hand clothing he'd worn, passed on from both of his older brothers.

I've come this far, he thought, and I haven't come this far to stop now.

The next summer, there were more trees to plant, more clients to move, and more fires to put out. Debbie kept the phones going, busy booking people, pilots, and figuring out the logistics.

Through the grapevine, George stopped hearing about how Far Away Helicopters would destroy him. Instead, he started hearing about how unhappy Willy was that George was in town. This wasn't a surprise – Willy made no secret of his unhappiness in many areas of life. The man had forsaken the sobriety he'd once enjoyed and now made a habit of keeping local barstools warm throughout the day and long into the evening, often becoming drunk in public. George knew Willy spoke badly about him around town, but it didn't matter anymore. If anything, it only made George work harder. Before they knew it, the Kelhams had the funds to buy more machines, and to hire full time staff. This time, the loans didn't feel so heavy.

The next step was obvious: Slave Lake Helicopters needed a hanger. For years, George and Debbie had kept their machines parked in a snowbank, covered with heaters to keep them from freezing, but the outfit was a pain for everyone involved. The problem was that the small town had no empty hangers available to rent. The Kelhams would have to build their own, but that remained a prohibitive expense. George and Debbie had anticipated the task, and had even bought steel to create a hanger, but only ended up with nowhere to keep it. The long beams lay alongside the helicopters, on the ground, outside.

The real trouble was that every time the company was in the position to build, a new expense would come up. Slave Lake was a difficult place to attract pilots to work in. It was out of the way, featured no night life, and offered few comforts, so Slave Lake Helicopters struggled to keep staff and paid high costs in their pilot's hotel room living expenses. George knew they'd have better luck hiring and keeping quality pilots if they invested in a crew house or two where their out-of-town pilots could stay, and so the purchase of a local crew house took priority.

But one day, around five years after Slave Lake Helicopters opened, a surprising opportunity came up. Sawridge Aviation, a fixed wing company, moved out of Slave Lake and left behind an empty hanger. A lawyer from the Sawridge band had already called both Bravo and Far Away Helicopters in hopes to fill the space, and now asked if George would like to lease it.

George knew Bravo didn't need it. He imagined Bravo wouldn't last much longer in town, not with the work Slave Lake Helicopters had taken from them and continued to attract. But George also knew that Willy Light would want to snap up the hanger immediately – just because he could, and especially if it meant keeping the space from George. Willy was so spiteful he'd keep the hanger empty just to keep George's machines out of it.

"Why are you calling me if you already called Willy Light?" George asked the man on the other end of the phone. "Sure, I want the space, but I can't afford to get into a bidding war with him."

The man chuckled. "Mr. Light refused to talk to me, Mr. Kelham. He said if the Sawridge Band wanted to lease him space, that the band chief would have to call and offer it, himself. Of course, Mr. Twin is far too busy for that."

George erupted in laughter. So, Willy was too important to speak to anyone but the Chief.

"Yes, we'll take the hanger, thank you," George said. It was hilarious. Thanks to Harmony Helicopter's poor treatment of staff, they had been upended as Slave Lake's primary helicopter company. Thanks to Bravo's shortsightedness, Slave Lake Helicopters had been born. Thanks to Far Away's arrogance, Slave Lake Helicopters was finally indoors. While George had created his local competitors, in creating them, he had created himself.

He and Debbie moved the company indoors with the help of the kids and Leroy and Andrea. The new hanger had no water, no sewer, no office, and no in-floor heating. Wash buckets would still have to be hauled by truck and trailer, but the machines were out of the snowbank and that made everyone happy.

Outside their door, George hung a welcome sign:

Slave Lake Helicopters. Home of the Good Guys.

~

Over the next few years, Slave Lake Helicopters stayed so busy in spring, summer, and fall that slowing down for the cold season was no burden at all. For the first time, George and Debbie were finally able to enjoy the life they had built together. They went to concerts and hockey games, and, together, the Kelham family travelled to distant places like Japan, Honduras, Australia, Dubai, England, and South Africa.

But joy was mixed with tragedy. About a year after Charlie left the Kelham's care and returned home to Thailand, Debbie received news that Charlie had died in a terrible car accident. The whole Kelham family grieved for the loss of their friend. Adding to the grief only months after, Charlie's family suffered a second major blow when his mother suddenly passed away.

"We should make plans to visit Charlie's father," George told Debbie and the boys. He had known how important friendship was when times in his own life were hard. Within a few weeks, the Kelhams visited Thailand to pay their respects to their grieving friend.

The loss was a reminder of how short and unpredictable life is, and how fleeting the moments spent with one's children are. The Kelhams had moved onto an acreage on the outskirts of town by that point, but Boyd and Taylor were growing up quickly and seemed to spend less time at home every day. They were busy with friends and part time jobs and girls and sports. George often worried about his children. Boyd, on the cusp of adulthood, seemed somewhat anchorless and sometimes drank in a way that reminded George of his own father's demons. He hated to think his firstborn might experience the misery of addiction. And young Taylor had gone through a period in which he planned on being a police officer but couldn't understand George's disapproval of his son intentionally putting himself in harm's way. George wanted better for him; he himself had taken dangerous risks with the hope his sons would never have to. But all conversations about these thoughts quickly turned into arguments.

"Fine!" Taylor would say. "If you don't want me to be a cop, I'll become a pilot!"

"Fine!" George would snap back, determined Taylor would make his own way in the world. "But you can log your hours elsewhere, because you won't get them from me!"

But before George knew it, it was time for the boys to fly the nest. Boyd graduated high school in 2002 and moved to Edmonton. Taylor achieved his diploma in 2005 and started a career in the oil and gas industry, making a home for himself in Calgary. George's worrying over his sons didn't stop just because they were adults, but he tried not to dwell on such things. He had faith in his children. They would carve out a life all their own, just as he himself had done.

In 2004, George celebrated his fiftieth birthday. Debbie organized a trip to South Africa as a present and scheduled an opportunity for him to "go supersonic." This had long been a bucket-list wish of George's. No stranger to flight, he longed to break through technological barriers and experience the ionosphere. Of course, Debbie had found just the people to make that wish come true.

The company was called Thunder City. The idea was that the pilot of Thunder City's sophisticated aircraft would fly George out over international waters where there were no laws prohibiting commercial supersonic flights. Then, George and the pilot would go up into the ionosphere, were they would race through the atmosphere: *supersonic.*

Debbie and George rented a car in Cape Town and drove through wine orchards alongside the sapphire coast. The drive and its view were beautiful and seemed to almost be the pinnacle of luxury. However, when the couple arrived at Thunder City, there was bad news. The aircraft was undergoing maintenance, and there was no way George would be able to go supersonic that day, or any time during the remainder of their stay in Africa.

George was disappointed but refused to let the unfortunate turn of events ruin the trip for his wife. Instead, they returned to Cape Town, and George suggested they take a boat tour to Robben Island, where South Africa's president, Nelson Mandela, had once been famously imprisoned.

George and Debbie both admired the formidable visionary. In their lifetime, Mandela had led a country to greater equality and full suffrage without condoning violence. He had achieved for his nation what had not been achieved in Rhodesia Zimbabwe, but no one could argue he hadn't paid that price with his own life. Nelson Mandela had spent twenty-seven years behind bars, all for the crime of insisting that all peoples were equal.

The feeling was somber on Robben Island. The tiny room – perhaps only eight feet by six feet – was exactly how President Mandela had left it, and George's own body hurt to imagine the large man cramped for so many years in such a small space. The cell was primitive. The floor and the walls were made of cold concrete and there was only a small window, the sunlight blocked by bars. It was cold, and damp, and there was a thin mattress on a bed that might as well have been built for a child. There was a small cup and spoon on the nightstand. Beyond the room's window was the work yard, where Mandela had spent thousands of torturous days under a hot sun, crushing rocks into gravel with nothing but a hammer and his own strength.

Standing in this space, knowing what that man had gone through and what he had later accomplished, George didn't think one bit about going supersonic. There would be time for that another day. On this day, there was nowhere more powerful, or more meaningful, to be.

~

On the business front, opportunities arose in several surprising directions. Slave Lake Helicopters became solidly established as the local forestry and lumber industries' first choice in flying, and their machines were so often in the sky they were even able to retire one, selling it to a new friend. Don Murray was a friendly, upstanding, and generous man who had built enormous financial success in the oilfield and wanted to purchase the Kelham's EC-120B for private use. After the sale of the machine, George and Debbie were further able to improve their fleet.

Andrea had also discovered alternate sources of revenue. First, she developed and gained approval for a program that would deliver power linesmen directly from the helicopter onto steel structures to perform maintenance. The idea was to outfit helicopters with a small platform that linesmen could step onto before climbing onto the pole, so that they could be delivered to problem lines without the need of boom trucks. This would reduce the power company's footprint on farming fields and save them billable hours.

The Kelhams hired an aeronautical engineer to develop the platform, which was bolted onto the craft, and with Andrea's help, the invention became fully certified. Slave Lake Helicopters called the program ASTEP and power companies such as ATCO loved it so much that from the moment of its certification onward there would always be a waiting list of power liners eager to complete their work without ever touching the ground.

Next, Andrea developed and gained approval for a helicopter program called Human External Cargo. This program was the first to be implemented in Fire Suppression efforts in Alberta.

Through this success, the Kelhams were able to purchase more machines and hired pilots to man them. They bought two more crew houses for the visiting pilots to stay in and took out loans with numbers so large it took Debbie's breath away, but they always did their homework first. They could pay back their debts with many returns over.

In other areas, Slave Lake Helicopters pioneered a partnership with a telecommunications company to put up fiber optics for Indigenous Reserves up north. And, later, the Kelhams received official commendation from the Ministry of Environment for their work in transporting seedling for planting. Slave Lake Helicopters would go on to help plant over 152 million trees, assisting in preserving the beauty of Northern Alberta's Boreal Forest.

One day, a man with a cane arrived at the hanger. At the office's front desk, he asked Debbie if she would call George in to talk.

"One moment," Debbie smiled, curious as to what the visitor needed. She picked up the phone and called her husband.

Moments later, George appeared. "I'm George Kelham," he said, hand outstretched. "How can I help you?

The man shook George's hand in return. "You already have," he said. He gestured to his leg, which George realized was artificial.

"I was in an accident years ago, when a load of pipe unloaded off my truck onto my leg. I just wanted to drop by and shake the hand of the man who helped save my life."

George was touched. Business was not only good; it mattered.

~

In 2007, Debbie crunched numbers and called a team meeting in the leaky, barebones shop in which they had built such wealth. Slave Lake Helicopters was finally in the position to build the hanger of their dreams.

It was time for the real fun to begin.

The year the hanger was built was so busy and exciting that it took George right back to those first years of flying with Leroy and Andrea, when business felt like a high-stakes computer game with points to earn and levels to pass. George and Debbie had to ask the bank for a loan and there were many hurdles to cross. Funds were released for every progress payment on the development of the hanger, and any little obstacle in construction could stop the payment and prevent George from paying his contractors. It was a stressful time, but failure was not an option – it never had been.

They hired an architect to make sure plans were most efficiently developed, but George and Debbie stayed closely involved in the project. George would fly all day and then land on the other side of the runway and walk across to watch the progress on the hanger. Before they knew it, the project was complete, and then business really seemed to flow. There was no more time wasted on transporting water, no more generators pumping electricity for space heaters. The machines were easy to care for onsite, and the pilots were happy. Andrea, with the help of her husband Phillip, wrote the Maintenance Control Manual and the Maintenance Procedure Manual. The Kelhams were able to hire their first Director of Maintenance, Peter Betz, who, at only twenty-five, was one of the youngest (if not the youngest) directors ever approved to work in Canada. George and Debbie also hired Aircraft Maintenance Engineers. It seemed their dream was complete. Their business had been built.

It was perhaps time for new dreams, George thought. Slave Lake Helicopters wasn't going anywhere and would provide for George and Debbie and each of its employees without fail. George knew he would never truly retire, but he was ready to live a life of more than just work.

A home-away-from-home: that was the next big goal for George and Debbie. They had friends who owned property in Arizona, USA, and they felt the pull of the warmer climate. George also loved the idea of making good on that goal he'd had as a young man, walking down the vast, sun-dappled Ohio turnpike. He was destined for America, though he hadn't realized it would take him so long to get there.

The only trouble was that Debbie and George couldn't find a house in Phoenix that they loved. It wasn't for lack of trying, but the Kelhams simply weren't attracted to any of the expensive retirement communities that seemed to move at slow paces, their main activities related to golf. Then one day, George realized they were looking in the wrong city.

"Debbie," he said, "I don't even play golf."

"Well," she said, "What do you like to do?"

Concerts, they both knew. The couple loved going to concerts, sports events, trying new restaurants, and taking in shows. And George was still an adrenaline junkie: he wanted to drive fast cars and test fly new aircraft.

"Vegas," he said, suddenly remembering the glamour and excitement of Fremont Street and the feeling of never wanting to leave.

"Yes!" Debbie agreed. "Vegas would be a lot of fun."

Once they realized what city their home was meant to be in, it took no time at all to find their perfect place.

On the way to their new home, however, there was a funny moment. By this time in their marriage, George and Debbie had achieved an enviable standard of financial success, and while they chose to continue living modestly in their day-to-day lives, they happily splurged in other areas, such as flying first class. However, as they waited in queue to board the plane that would take them to Las Vegas, a ticket agent called the couple out of the line-up.

"I think you might be in the wrong line," she told George and Debbie, her eyes scanning their simple, non-designer clothes.

"We're in the right line, Miss," George told the ticket agent, showing her his ticket.

The agent's face turned bright red. She apologized vehemently, realizing the social blunder she had just committed in assuming the Kelhams could not afford first class travel because of their outfits' lack of frivolous adornment. The Kelhams accepted her apology and she left them alone, but the whole situation was humorous to George. It reminded him of the classism of Rhodesia and, despite the old systems that had fallen away since that time, George knew there would always be judgement, assumptions, and discrimination in the world. George would never have a family name that inspired awe, an elite lineage, or wealth that had amassed through generations of men who built fortunes on the backs of others, but George didn't care. His fortune was built through the sweat of his own brow, and success was much better this way.

George was not born into wealth and did not marry into it. He and Debbie had accomplished the impossible, all on their own, and they hadn't squandered a minute of it. This truth was so remarkable it sometimes brought a tear to George's eye.

In that same year, Willy Light died.

WILDFIRE
2011

The spring of 2011 was so hot and dry that branches snapped in the wind and grass turned brown and was blown off parched, cracked earth where only the year before it had been lush and green. In May, Andrea visited the Kelham home before returning to Quebec for her days off, and she and Debbie were sitting outside when Andrea suddenly had a terrible thought that hit her like cold water.

"Jesus, Debbie," Andrea said. "Look at how dry those trees are." She gestured to the forest which lay on the other side of the highway that separated them from the Kelham's home. "One spark, and this whole forest would go up in flames."

Debbie nodded. It was very hot and dry and windy, as May often was. But Andrea was upset. She couldn't shake the feeling of risk, as though her thought had not just been a thought, but a premonition. The next day, Andrea went home as planned, the whole way feeling a sense of unshakable concern.

By that time, the Kelham boys had returned home from their respective cities as grown men. Boyd managed a local restaurant and had started seeing a local girl named Lexi, and Taylor had returned to Slave Lake engaged to a pharmacist named Kristine, who soon found work at a local pharmacy. Plummeting oil prices had made life in Calgary undesirable, but Taylor had kept one foot in the oil industry, finding greater opportunities for employment in the north. He had also followed in his parent's entrepreneurial journey, buying a local athletic goods store called Sports First from a retiring merchant. George and Debbie were thrilled to have their sons once again living nearby, and both adored their future daughter-in-law. Kristine was lovely and grounded and smart, and the love and respect Taylor and Kristine showed one another reminded George and Debbie of earlier versions of themselves. When Debbie was with Kristine, she remembered the immediate affection Mrs. Mayo had shown her, and entirely understood the love between a mother and her daughter-in-law.

On May 14th, 2011, George was flying for Alberta Forestry. His assignment was to survey the forest around Slave Lake and deliver firefighters and fire suppression tools where needed. It was fire season, and forest fires were usual in northern Alberta, so George didn't anticipate anything out of the ordinary. However, as he began his flight back to the hanger, it was obvious a massive problem had occurred.

Within hours, the whole town knew of the fire that raged just a few kilometers from the boundaries of their community, and conflicting reports of the situation spread even more rapidly over social media. Taylor phoned his father, knowing George's report would be free of all hysterics. George was always as cool as a cucumber in times of crisis. For years, when reflecting upon their father's lack of spontaneous emotion, Boyd and Taylor had remembered one seemingly unremarkable day: George had been trying to locate something in a shed with his back turned to the boys when one of them had knocked a steel barrel out of the back of the family's pick-up truck. When the barrel had hit the asphalt driveway, it had caused a terrible, thunderous boom that made both boys jump, even as they had watched it fall. But George's shoulders hadn't even flinched. He found the yard tool he had been searching for and calmly turned around to ask them, "What was that?"

"So, what can you see up there?" Taylor asked George. "Is this thing going to be a big deal?"

"Well…" George said, "it's not good." He'd seen how massive the fire was, and how close it was to the southeast quadrant of town.

"Not good?" Taylor was surprised, having had fully expected to hear words of reassurance.

"I mean… don't worry yet," George said.

"Should we be ready to leave?" Taylor asked, trying to get a straight answer.

"We should keep an eye on it," was all George would say.

That night, all members of the Kelham family struggled to sleep, their eyes and throats sore with the smokey air that had permeated into their homes and made their chests tight with worry. The next day the situation didn't look any better. The wind had picked up, and the whole town was on alert.

George took to the sky as soon as he could, hoping to brief himself on the situation. There was an eeriness to the energy in town. It was as if all residents were in a state of pause, just waiting for the other shoe to drop. It reminded George of the St. George's Day ambush so very long ago, when for a moment it seemed the whole world was silenced.

Flying to the southeast, George felt as though a stone had dropped in his gut. The view below his craft was orange, dark, and grained with smoke. Fire blew in long-stretching flames over everything in its path below. George had never seen fire burn uncontrolled so close to town.

"It's passed the lumber mills," George phoned Taylor, after he landed. "This is not good."

Taylor was chilled by the sound of his father's concern.

"Well, that's still pretty far from town, though, right?" Taylor said. "They still have time to get it under control."

George thought about the high winds, and the matchbox-dry forest. "I don't know," he said. "This one, I don't know."

Within two hours, he called his son a second time.

"This thing is coming for us," he told Taylor, yelling over the howl of the wind, almost in disbelief of his own words. "It's coming, and it's not slowing down."

"Holy shit," Taylor said.

George told Taylor to locate Boyd and their mother, and then to get as much of George's long-curated hockey collection out of the house as possible. George and Debbie's home was directly in the fire's path.

Taylor hung up the phone, not knowing where to start. There was so much to do to protect himself and Kristine, his mother and brother, his friends and his home and his belongings.

Boyd, his father had said.

Yes, Taylor thought, he would need his brother's help.

Hearing the seriousness in Taylor's voice, Boyd told his brother to leave his parents' belongings to him.

"Bring Kristine home from the pharmacy," Boyd said. "You two pack and I'll get a few guys to help me with dad's stuff. We'll meet at the hanger, okay?"

Respectively, they went to work. Taylor phoned his fiancé and told her to come home and pack, and poor Kristine had rushed around the house with so much on her mind – the items to pack, the patients she'd need to ensure had their medication, the fear of the disaster – that she fainted right before Taylor's eyes, having remembered to do just about everything except to breathe. Boyd found Debbie at home making a large pot of chili and roasting two chickens to feed Slave Lake Helicopters' employees.

"We have to pack up dad's stuff, mom," he explained when she asked why he had brought home friends.

"Why?" Debbie had asked.

"Dad wants us too. This fire is getting bad."

Debbie didn't believe it but just then, the phone rang. It was George.

"Debbie," George said. "I need you to find the kids and get out of town."

Debbie's blood ran cold. George was never panicked, and she knew him to act slowly, calmly, methodically; if he was telling her the situation was dangerous, it was very, very dangerous.

"George, Boyd is here, and he's been in contact with Taylor. We will stay together and pack, but we are not separating," she said. "We are not leaving you. Just tell me, where should we go?"

George knew there was no arguing with Debbie. He thought carefully while looking down on the town, the walls of wildfire rapidly approaching, spurred on by high winds.

"We'll meet at the hanger," he said before hanging up. "For now, just go to the hanger."

"Boyd," Debbie said, interrupting her son and his friends as they packed jerseys and autographed posters. "This is very serious. Have you spoken to Lexi yet?"

Boyd shook his head. "She didn't answer her phone. I'll drop by her place when we're done here."

Debbie shook her head. She knew her husband wouldn't upset her with such a worrying request unless things were critical.

"No," Debbie insisted. "You might not have time if you wait any longer. You need to find Lexi now."

The men in Debbie's home looked at one another, all suddenly very aware of the many loved ones in their lives.

"Okay, Mom," Boyd said, before thanking his friends for their help.

When Boyd started his vehicle, he was shocked to hear the urgent evacuation notice playing on repeat on the town's radio, and to see the line of traffic that snaked through the highway. Fire burned on either side of the road and the back of his parents' tree line was already ablaze. He was filled with terror, and punched Lexi's number into his phone while driving. Still no answer.

"Shit!"

Boyd drove in the opposite direction from the other vehicles, the townsfolk fleeing the southeast as he headed into it. Lexi lived by the local high school but when Boyd reached the neighborhood, he saw a barricade had been set up by police, blocking his route. There were no members keeping watch, so Boyd had no choice but to go right through, his vehicle lurching as he collided with and sped over the splintered barrier.

When he reached Lexi's home, he pounded on her door to no answer, all the while still trying to phone her. Her vehicle was in the driveway, so he knew she had to be home. All around him people were panicking, throwing items into vehicles, and shouting at neighbors for help. Finally, Boyd couldn't wait any longer. He kicked the door in and entered Lexi's home, rushing to the sound of a hum in the basement. Downstairs, he found Lexi breathless, running on a treadmill. Her eyes widened in surprise as she removed her headphones from her ears.

"Hey!" She exclaimed. "What's up?"

Back at the Kelham house, Debbie noticed the fire in the backyard. Dark smoke seeped into the house and her mind flashed back twenty years to the day of the flood, when she and her babies watched water rise higher and higher in their home until it was too late to evacuate on their own.

Mom, Debbie silently prayed, we need your help today.

Debbie had already grabbed the chili off the stove and packed it into the back of her car with the hot chickens. Now she ran around the house as quickly as she could, trying to decide what was to be saved. She knew they could lose almost everything and still be alright. The flood had taught her that. Even after she and George had lost every photograph and memento they shared, they had still had one another. In the end, Debbie snatched costume jewelry her mother had left to her – worthless in value, but high in sentimentality. Then, Debbie drove to the hanger.

The air was so thick she couldn't see what side of the road she drove on, or if anyone shared it with her. Once at the hanger, Debbie thanked God for every person that entered the door: Taylor was safe, Kristine was safe, Boyd was safe, and his girlfriend, too. Soon all the crew members and their families were there, and together they all hugged one another and cried and crowded around the small radio, listening for any news at all. The power was out at the hanger, and the air was so hot and smokey Debbie wondered how much longer they could stand it. Then the phone rang: George.

"An excavator has hit a powerline just down the road from you," George said. "It's caused a second fire that's burning near the runway and working its way to the hanger. You guys are sitting on too much jet fuel. You've got to get out of there. Go to the lake," he said.

But just as he said it, George changed his mind. "No. Never mind. I can't get my craft out there. Go to the baseball diamonds, instead. I'll pick you up. Take water with you. And Debbie? I think our house is on fire."

In the air, George was filled with a dread he'd never known before. For him, this was worse than Rwanda, worse than the Bush War. In every danger George had been in, it had only been his life he felt at risk. Now, it was Debbie and his children and almost everyone he knew. From his view, the town he and his family had loved for so long looked like a war zone. It was hell on earth below him: explosions peppering gas stations and car dealerships, vehicles fleeing between walls of rushing flame. There was a row of townhouses on thirteenth street that had become a raging inferno, burning so violently George could barely take his eyes away, but felt sick watching. He had passed over the fire near his own home and had seen flames licking up a wall, and now the flames burned so rapidly in that area he was sure the whole house was lost, too. George had never felt so powerless in his life. From the sky, he could only hope no lives had been lost.

Debbie started getting things ready to leave right away. She flew through the office, packing anything the business might need if the hanger was to burn. She detached computer hard drives from their screens and threw them into the backseat of her vehicle, which she now saw was covered in chili from the uncovered pot she had hoped to have saved.

"Pack the logbooks," Peter Betz insisted. "We can't replace

the journey books!"

The whole team fled to the baseball diamonds. Everything was chaos, the world thunderous with honking horns, distant explosions – gas tanks, they'd later learn – and the crying and shouting of fleeing residents. Everything around them was on fire: trees, fences, homes, vehicles.

But they were safe in the wide-open expanse of the baseball diamonds. George landed his craft on the field and informed Debbie that he'd been wrong: their house still stood. A jerrycan near an exterior wall had been on fire, but firefighters had put out the blaze. He asked Taylor and Boyd if they wanted to survey the damage, but Boyd opted to stay on the ground with Lexi, who needed his support.

In the sky, Taylor was amazed by the damage. More houses caught aflame with every lap he and George flew around the town, until the entire southeast quadrant was decimated.

The Slave Lake Helicopter team stayed at the diamonds until police advised they take shelter at the local Honda Shop, where they could get out of the smokey air. As hazy light became dusk, George reported that the fire near their home had been put out – or, rather, had burnt itself out after ravaging all fuel sources. It was safe to spend the night at home.

The blackened town was nearly empty as the Kelhams and their friends drove to George and Debbie's home. The absence of the thousands who had left made the town feel ghostly in its emptiness. George, Debbie, and the Slave Lake Helicopters team were given special permission to stay in town as essential workers.

That night, eighteen people rested in beds, couches, and on the floors of the Kelham home. They ate crackers and cheese because there was nothing else. There was no water at the house and the phonelines were down, but there was electricity. George and Leroy flew helicopters and fuel out of the endangered hanger and parked their crafts in the ditch by the highway near the Kelham home.

For a second night, nobody slept. Emotions were high and people swung between erupting with relieved laughter in gratitude for the preservation of their lives and sobbing with grief for their homes and the community. Debbie tossed and turned, wondering how she'd feed all the people who had become indefinite surprise house guests. Grocery stores and restaurants had been locked up and abandoned. Luckily, the very next day, friends called with the very answer to that question.

"Do you have food?" Don Murray asked Debbie.

"We need all the food you can send us!" Debbie told him.

A day later, George secured special permission for Don and his wife Linda to land his jet at Slave Lake airport, and they saw with delight that the plane was filled floor to ceiling with food.

George and Leroy continued to work the fire. George was assigned to an assessment machine and a forestry officer from Valleyview named Emile Desnoyers travelled with George, surveying the damaged community with him for days on end. Debbie and Boyd helped feed everyone they could, sending those who didn't need to stay in Slave Lake to Edmonton. Then they went out into the community and did what they could to support firefighters. Taylor and Kristine manned the pharmacy, which she couldn't abandon as an essential health worker. They made sure everyone who had stayed behind to aide recovery had access to the medicine they needed to stay healthy while fighting the fire.

Sadly, one life was lost to the fire. George and Emile were flying over Canyon Creek to fill a hanging water bucket one smokey morning when George saw something shining in the lake and soon realized it was a crashed Bell 212. The men called the loss in, and dispatch sent authorities to investigate. The helicopter pilot was not known to George. His name was Jean-Luc Deba, and he left behind a wife and two children.

By the middle of May, headway in the fight had been made, but much had been lost. Some of Slave Lake's residents returned to scorched earth where their homes had stood, and some decided not to return at all. Eventually, the chain of pharmacies Kristine worked for asked her to move to a pharmacy in Edmonton to help displaced Slave Lake residents there. The mood throughout the province was emotional, and often tense. At West Edmonton Mall, where it seemed the whole town had congregated for weeks following the fires, Taylor came close to a fistfight with a man who had accused George of dropping buckets of water on his house instead of fighting the fire as commanded by the fire department. There was no truth to the tale; physics itself demanded that water dropped from the chopper's bucket onto the house would have flattened the structure, but people believed what their pain convinced them of.

Many houses were totaled and rebuilt. The Kelham's insurance provider wanted to knock down their home because of the roof's fire damage, the singed siding, and the smoke-damaged interior, but Debbie fought to save and restore the home herself.

The summer that followed was one spent mourning, rebuilding, and planting trees, but celebration came in the winter.

~

On New Year's Eve, 2011, Taylor married Kristine the same day George celebrated his fifty-sixth birthday.

The occasion was a highlight of the Kelhams' lives in so many ways. It was the beautiful union of Taylor and the wonderful Kristine, but it was also the only time all the nieces and nephews of George and Debbie's combined family were together. One of Marion's sons had died early in life, but her remaining child, Johnny, attended the wedding with his wife, and Debbie was pleased to see that the little boy she'd met as an infant during that terrible lunch had grown into such a thoughtful and charming young man. All of Debbie's family was there, and George's brother Rob attended with his daughter. Rob's marriage to Anne had not survived, but it had produced a child who had grown into a beautiful young woman named Julie.

The affair was a bit chaotic, as most good things are. Hotel servers failed to show up for the wedding on account of it being held on New Year's Eve, but the bride and groom didn't seem to care. No member of the Kelham family was afraid of hard work, and they took care of putting out and cleaning up the catered meals themselves. The Canadian hockey team was playing USA that night, and George and Boyd snuck a television into the reception so Taylor could watch the game with them without Kristine noticing. Of course, Kristine noticed, and she didn't mind the fellas having their fun. It was all meant to be a simple joke, but Canada won and the whole wedding party celebrated the victory.

EPILOGUE

HOME
Forever After

In the years following the fire, time seemed to speed up. Many of the children George and Debbie once knew now had their own families. Grant Penalton's little girls, who George had always bought packets of Smarties for, were now grandmothers themselves; the eldest daughter married to a linesman, just like her dad. It seemed there was proof of time's passage everywhere George looked.

One day, George was tasked with flying a helicopter attack crew with their leader, Mike Turcotte, north to Peerless Lake, where a minor grass fire had been set by children playing with matches. The fire turned out to be so small that use of George's helicopter bucket wasn't necessary, so George simply parked the machine. While waiting, he was suddenly struck with the urge to knock on a house door and to ask about a child he often wondered about – the little baby he'd flown to hospital, riddled with pellets.

A woman answered the door of that home, and when George described what had happened so many years ago, she knew exactly who he was talking about.

"Yes, that's Greg!" the woman exclaimed.

"So, he's okay?"

The woman smiled. "Oh, yeah! He still has some scars on his face from that night, but he's okay. Good thing you got him to hospital, eh?"

Reconnecting with old friends became an important part of

George's life. Somehow, George found DC Sueing's phone number – the man with whom he had washed dishes in Detroit – and George flew down to the city to watch a hockey game and take DC and his wife out for dinner. It was a lovely evening, but the next time he called, the phone was answered by one of DC's grandchildren. Their grandpa had passed away, and George was sorry to have lost a friend.

Sean Griffith had also passed. He had fought a long battle with cancer. But before he passed, George and Debbie had met up with him in London where he happened to be living with George's former fiancé, Eva O'Kelly. As big as the world seemed, they were all Umtali kids at heart. Life, in some way, would always feel like a small village.

On that same trip, George and Debbie had visited clever Sarah, George's English cousin. Sarah had made a good life for herself working for the luxury department store Harvey Nichols in London, where she planned and purchased new clothing lines in Paris every fashion week. Aunt Ena had passed away, but their absence didn't at all affect the strong affection George had for the wonderful women.

Sabastian, Neil, and Clive were still in South Africa and, while visiting them during George's fiftieth birthday visit, George and Debbie also reconnected with Hannah's sister, who reported her happy marriage. He didn't know whatever happened to Pixie, Steven May, or Priscilla, but, wherever they were, George wished them all the best life had to offer. Jimmy Hynes, he discovered with much delight – the wee boy with whom George had so loved lighting the world up on Guy Fawkes Day – had made a home for himself in Santa Clarita, California. Fittingly, he had become an explosives engineer.

And, of course, the Kelham family had grown up, too.

In May 2015, Taylor and Kristine called Debbie in the middle of the night to let them know their first grandchild would enter the world in a matter of hours. Debbie was so excited she ran around the house just as frantically as she had the day of the fire, packing things on short notice, unsure of what she'd need – only this time, she felt as though she might explode with joy. Coming to her senses, she stopped, found her glasses, and called George, who was away purchasing another helicopter.

"Hello, wife," George said, answering the phone. In all the years they'd been together, Debbie had so seldom struggled to get a hold him when she needed too, regardless of time differences or whether it was day or night.

"Our baby is coming," she laughed.

George cheered.

"Oh, George," she said. "I wish you were here."

On the other side of the world, George smiled.

"I will be soon," he promised. "Then no one will be able to pull me away from that kid."

George was true to his word. Baby Tucker became George's best friend and most frequent flying buddy, never denied candy or a cuddle from grandpa. The sight of that boy never failed to stop George in his tracks.

The next year, 2016, was one of big life transitions for George's sons. Boyd, having been raised by a man who had demonstrated how to live a happy and successful life free of drugs and alcohol, achieved sobriety. George watched with relief and happiness as joy and health flooded back into his son, and he saw there was no more risk of Boyd living in the same pain John Leonard had chosen. George's son had waged his own war and this one had been winnable. Boyd became a man of resilience and integrity, earning a reputation of kindness and reliability within his community.

Following in the footsteps of his parents, Boyd nurtured and cared for his relationships so deeply that friends became family to the young man. His money, which once had been in constant short supply, was soon abundant enough to spend on hockey games, playing golf, and taking his dad to concerts all around the world, once logging thirty-two live performance acts in one single calendar year. At those venues, the Kelham men drank lemonades or cokes. They banged their clear heads to hard music, enjoying experiences that neither father nor son would ever forget.

In Taylor's world, there was a different rock bottom. Oil prices plummeted and Alberta again entered recession. George had always worried the oil and gas industry wouldn't be good to his son and it seemed his concern proved correct. There was no more work to be done in the oilfield – or so it seemed, anyway – and a reliable livelihood in the industry likely wouldn't be a possibility again for many years. Adding to Taylor's unique misfortune were two facts: Slave Lake's economy heavily relied on the success of the oilfield, and, because of its failure, Taylor didn't only lose work, but residents of the community had no extra funds to support his store, Sports First. Secondly, major global conglomerates like Amazon had begun delivering cheap goods to the town, and it became a trend for those who had the money to spend on sports equipment to spend it online. Seeking greater security for his family, and certain of his ability to aide his parents' business operations, Taylor proposed that he come aboard as a staff member at Slave Lake Helicopters.

At first, George gave the idea a hard pass. He wanted his son to take on the world, just as he had. He wanted Taylor to fight for what he needed as George knew no other way of developing skill. But Taylor did fight. He fought against George's refusal. Taylor showed up for work at Slave Lake Helicopters every day and worked for free until George took notice of his son's success in the role. Taylor noticed the large workloads Debbie was carrying, managing the office all by herself, and brought in an administrative assistant named Tammy who George and Debbie would come to trust as their Gal Friday, one of their closest confidants. After months of noticing his son streamline processes, ease his mother's burden, and bring on new accounts, George was proud to welcome his son into the business.

Then it was George's turn for a personal crisis. In March of 2020, George visited a dentist for a checkup.

"Have you had the lump on your lip looked at?" the dentist asked. George was surprised. He had barely noticed it. Scar tissue had formed around the same area the doctor was concerned about from the frozen screw he'd pulled from his mouth decades earlier, and so no bump was especially noticeable.

The dentist sent George for a biopsy, and the results came back positive.

George had cancer of the mouth, an outcome the doctor believed was likely because of constant sun on his face; a result of his lifetime of flying and his experiences in Africa.

The days leading up to George's surgery were reflective. Cancer had loomed heavily in George's life, first taking his mother and then his old army buddy, Sean Griffith. George had lived a full life, that was true. He had travelled the whole world, seen amazing sights, and met incredible people. He had built something from nothing many times over. He had found family, and he had provided for them. He had found a true home.

No longer one of three young boys who had to jockey for the best bed in a DIY bedroom tacked onto the back of a tiny house on Rudland Avenue, George had truly known success. He had come from nothing, starting life as the son of a fitter-and-turner and a photographer's bookkeeper – working class people with no other option but to outfit their youngest in third-hand clothes. Somehow, through a whirlwind of hard work, good luck, and almost constant upsets, George had amassed for himself and his family a wealth that still seemed incomprehensible. He had three beautiful homes: one in Canada, one in Vegas, a third in Honduras. He owned additional properties: staff houses that he sometimes forgot he had even bought. His machines – now counting ten, and none of them rented – were no longer housed outdoors but kept in one of three state-of-the-art hangers. His business directly supported more than twenty families, and his company held over ten vehicles. George and Debbie had also hired and befriended a landscaper nicknamed Ten-Bob-Block, who George knew to be perhaps the kindest, most humble man in the world.

Although once a boy routinely abused, insulted, and sneered at by teachers, commanding officers, and employers, George had made himself into a man who received letters of commendation from government ministries, and thanks from those whose lives he had helped save.

And perhaps most importantly, he had kept promises to himself. He'd promised he'd make life in Slave Lake work, and so he'd created a fruitful business. He'd walked the Ohio turnpike and promised himself that America was his destiny, and so he had bought a home in Vegas. He'd washed dishes a stone's throw away from Detroit and promised himself he'd someday go to any concert he wanted to, and George had made that dream come true, too. He'd watched hockey at dive bars in Fox Creek and imagined what it might be like to be in the arena until he and Deb became season ticket holders for thirty years in a row. As a teen, George had seen that Lamborghini outside Pop's Café in Umtali and had promised himself then he'd one day have one. Neil and Clive might not have believed him at the time, but George had made that promise come true, too – again, and again, and again, and again, relishing the ownership of five luxury cars. He'd many times known the hunger of having too few coins in his pocket for even the cheapest of fast-food fare and had transformed his bank account to that of a millionaire. He hadn't yet gone supersonic, but he still held out hope that someday he would. Most importantly, he'd always wanted to become "that guy," the one who took to the sky. Then, he'd achieved a flying career spanning over forty-one years, logging over 17,000 flying hours.

There was only one promise George had made that he hadn't fulfilled. He remembered it clearly: if he made it out of Rwanda, he'd someday write a book.

As George got older, the idea of writing a book became more and more important to him. He knew he wouldn't be around forever, and yet there were so many stories he hadn't told his children, his friends, or even Debbie. Some stories were too hard to explain in light conversation. Some were so dark or too frightening to have shared with the boys as children. Many he had kept to himself because he feared souring the idea his young sons had of their father. He didn't want his children to associate the dangers of the world with him. All he had ever truly wanted was to be a safe place for them, to be part of their lives, and to protect them.

But the thought of his stories dying with him felt like a violation, an injustice. He began considering having them written down as he knew his skills didn't include wielding a pen. Jimmy Hynes's sister, Angela, had become a writer. She lived in Hollywood and had enjoyed success writing for television. George asked her if she could help him, but she insisted she was too old now, retired from the gig.

Still, it wasn't like George not to achieve a goal.

George went under the knife in May 2020, only two months after his dentist's appointment. It was successful and all the cancerous material was easily cut away, although the six-week recovery period was intensely painful. Still, George only missed a day of work, coming into the hanger around the same time the anesthesia wore off – until Debbie chased him home, that is.

It was about a month later when George fueled up his vehicle in Slave Lake to glance over the pumps and see a former wildlands firefighter he'd transported to and from wildfires about twenty years prior. The man was on the cusp of middle age and had left forestry to pursue urban firefighting and paramedicine in a city near Edmonton. George knew the firefighter had married a local girl and taken her with him to their new life, but that they often returned to Slave Lake to visit relatives.

"Hello, Fred," George said. "It's good to see you! How's the family?"

"Good," the firefighter said. "Really good! My wife just published her first book."

The wheels in George's head began turning. He smiled. Most of his life could be explained by resiliency, by hard work, and by courage, but some coincidences were inexplicable. Some he could only sum up to the person watching over him, the person who looked like she might always keep vigil from a departing train. Some things could only be thanks to a mother's love.

Thanks, mom.

"An author, you say," George laughed. "Think of that."

ACKNOWLEDGEMENTS

The overarching story of my life has been a quest: for home, for family, and for success. There is only one person in this world in whom I have found all these treasures. Debbie - my sweet, kind, gentle, patient, loving, wise, courageous wife - I don't know where or how to start with my thanks to you. Thank you for allowing me the freedom to reach for my dreams. Thank you for forgiving me my shortcomings. Thank you for believing in me when it felt like no one else did. Thank you for the family you so thoughtfully nurtured. Every good thing I have is owed to you. There has never existed a finer wife.

To my sons, Boyd and Taylor Kelham, your very existence brought two pivotal elements into my world. First: the joy of fatherhood. Thank you for making me your dad. Second: the drive to provide. You gave me the fuel to carry on, even when it felt like I had nothing left. I'm proud of you both. Never forget this: I am proud to be your father.

There is another young man in my life whom I love, but before penning a message to him, I must first express my appreciation to the child's excellent mother. Kristine, thank you for bringing our beautiful grandchild, Tucker, into the world. You have completed our family. We could not wish for a more wonderful daughter-in-law, and we love you.

Dear Tucker, as I imagine you reading this book as a young man, possibly long after I'm gone, there are tears in my eyes. My message to you is the hardest to write. But my message is this, and it always will be: I love you. You are the champ. You are one of the happiest parts of my life.

I must give thanks, too, for the many wonderful friendships I have been fortunate to experience. Thank you to the first friend I ever had, Jimmy Hynes. Thank you to those friends who became chosen family: Cathy and Bill Mayo, my surrogate parents, and my chosen brothers, Clive and Neil Mayo. Thank you all for accepting me and for taking care of my bride when I had little family to offer her. Thank you for allowing me to give Debbie a loving mother-in-law, one hell of a wedding party, and unwavering support throughout our lives. It is with great sadness that I acknowledge the death of a close friend. It is the world's loss, and certainly mine, that Clive Mayo passed during a heart operation in December 2021, as we neared the completion of this book. I wish he could have read it and I hope he knew how much I appreciated his friendship.

Thank you to Sebastian Wince, Sean Griffith, and Steven May, for teaching me how to be young. Thank you to my brother Robert, and to Anne, for the generosity you showed during my first years in Canada. Thank you to Wendy and Dennis Barton for friendship and encouragement in tough times, and to Rick and Joanne Forberg for the great laughs we've had together. To our excellent American neighbors, Jim and Sandra, thank you for always helping us feel at home away from home with your friendship. To one of the only people to make "George's Approved Cooking List", Marilyn Dawson, thank you for being such a wonderful friend to my wife. Richard Barrett, thank you for being a source of steady strength during the rough times we went through overseas. To my cousin Sarah and my dear Aunt Ena (may she rest in peace) – the first extended family I ever met – thank you for welcoming this long-lost nephew/cousin into your home with open arms, and for loading me up with more food than I could carry during those hungry London days.

A special thank you to the friends who were instrumental in the success of Slave Lake Helicopters: Leroy and Andrea. There are no better pilots or people than you two. The best days of my career were spent flying with you. Andrea, you have brought so much innovation to our company. Leroy, you may possibly be the nicest man on the planet, and I am so thankful for our 40 year friendship.

Of course, perhaps no person was as directly responsible for the success of Slave Lake Helicopters as the esteemed Bob Vanderwell, may he rest in peace. Through Bob, I learned pivotal lessons in what it takes to bring ones' business to the heights of success. Without Bob's faith and support in our operation, Slave Lake Helicopters would likely still be an unfulfilled dream. There has been no man more influential in my life.

Thank you also to the many amazing clients who have kept us high in the sky, including Vanderwell Contractors, Buchanan Lumber (now West Fraser Timber), and ATCO. Along the way we've forged relationships with so many good people who have put a lot of trust, time, and investment in our ability, our staff, and our equipment, and we couldn't be more grateful. I would also like to extend a small note of thanks to Mike Reno of *Vertical Magazine*, who played a part in bringing news of Slave Lake Helicopters' happenings to the readership of his excellent publication.

To a young man who didn't get to fulfill his dreams, my big brother Francis: thank you for showing me that the world could become mine for the taking, and that life could be exciting and bold. I hope I'm living fully enough for the both of us.

And thank you, Mom. Thank you for the love you gave and the prayers you prayed. While I am not a religious man, I have always believed in you, and I know the words you once uttered protect me still.

Finally, a thank you to every person – every teacher, every commander, every shortsighted employer – who ever told me I couldn't.

How sweet it is to prove you wrong.

APPENDIX

John Leonard and Louise Susannah on their wedding day, 1941

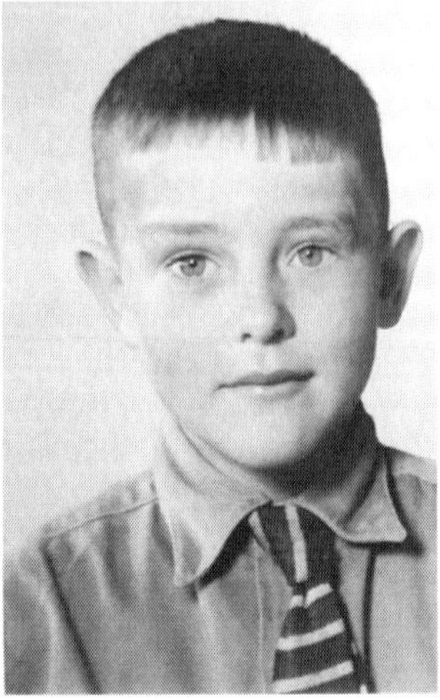

A Early School Portrait, 1964

Playtime in Umtali

Fifteen year old George and crew members work to erect power lines in Rhodesia

George and his Kawasaki 900cc, 1974

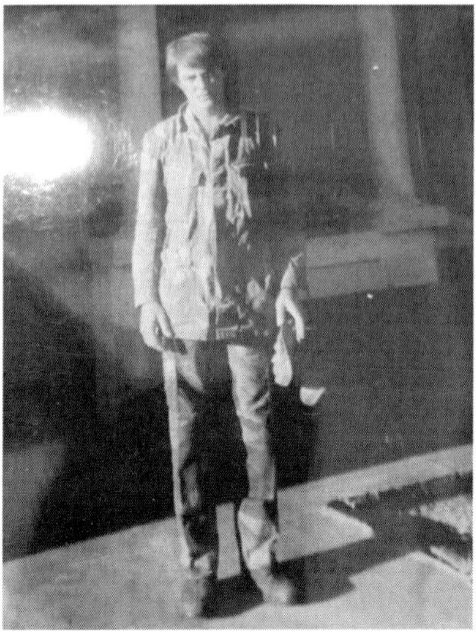

Seventeen year old George prepares to return to Llewellyn Barracks

The first of two trucks attacked in the St. George's Day Ambush

George at his sky-diving lessons in South Africa, 1974

Twenty year old George prepares to hitchhike across America, 1974

Sign reads:
"Rhodesia to Chicago"

Twenty-one year old George in London, England

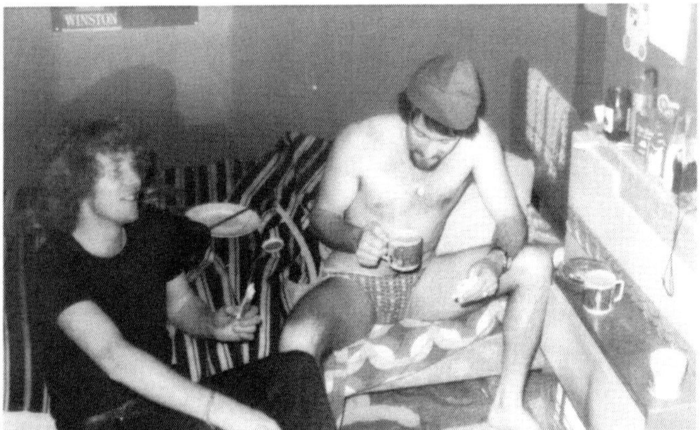

Steven May and George share a laugh in the lads' one room apartment

Power-lining in Canada proved to be a much different experience than in Africa.
Fox Creek, 1978

George and a crew member configure lines in northern Alberta.

George and Debbie, 1981

Pilot Kelham with his new son, Boyd

The young Kelham family depart by train from Johannesburg to Cape Town. South Africa, 1988

A Russian Mi26 helicopter.
Mogadishu, Somalia, 1994

A Somalian woman washes laundry beneath the shade of a Russian Mig.
Mogadishu Airport, Somalia, 1994

George attends a morning safety briefing on medivac training in Somalia, 1994

Rwanda
Richard and George document the morning after the downing of the Rwandan president's plane (smoke from the attack can be seen over the building, right.)

A Hutu insurgent patrols below George's hotel window during the Rwandan Genocide, 1994

George helps the Italian Explosive team destroy caches of weapons following civil war in Mozambique

The original Slave Lake HelicoptersTeam

Andrea and the ASTEP platform

Andrea, Debbie, and George on ASTEP's First Day of Operation

Leroy and young Boyd

Leroy and baby Tucker

The Kelham Family
Taylor, Tucker, Kristine, George, Debbie, Boyd

Three Generations
George, Taylor, and Tucker standing in front of a Slave Lake Helicopter

Supercross, Sam Boyd Stadium, 2019
George and Boyd strike a pose at one of many events shared.

Then...
(Young George in his beach buggy in front of his childhood home, 1971)

...and Now, Fifty Years Later
(Present day George with his Lamborghini in front of his Las Vegas home, 2021)

The author and editor at work (2020)

ABOUT THE AUTHOR

George Kelham is a former United Nations pilot who has lived and worked in such places as Rwanda, South Africa, Greenland, Mozambique, Tanzania, Botswana, Britain, The United States, Guatemala, Somalia, Canada, and the former nation of Rhodesia. He is president of the successful helicopter company, Slave Lake Helicopters, which he owns in partnership with his wife, Debbie. The Kelhams reside in both Northern Alberta, Canada, and in Southern Nevada, USA.

ABOUT THE EDITOR

K.H. Bickell is the recipient of the Georges Bugnet Fiction Award, the Howard O'Hagan Award for Short Story, the Writer's Guild of Alberta's Emerging Writer Award, the Alberta Views Fiction Prize, and the Voices of Motherhood Prize. Her novel, *Always Brave, Sometimes Kind*, was shortlisted for the national ReLit Award for Best Novel. She lives in Sherwood Park, Alberta, with her husband and daughters.

ABOUT THE FRONT COVER PHOTO

George stands next in front of the first helicopter he flew while working for Kenting Helicopters, Calgary, Alberta, 1980.

Made in United States
North Haven, CT
10 November 2026

22088874R00148